RAND

The Impact of Water Supply Reductions on San Joaquin Valley Agriculture During the 1986–1992 Drought

Larry L. Dale, Lloyd S. Dixon

Prepared for the
United States Environmental Protection Agency

PREFACE

In late 1994, California adopted more-stringent water quality standards for the San Francisco Bay/Delta and the San Joaquin and Sacramento River System and is currently deciding how the water-use reductions necessary to meet these goals should be split between agricultural and urban users.

There is substantial debate on what the effect of water supply reductions on agriculture might be. This report attempts to improve understanding of the likely effects by examining economic theory, past empirical work, new data on the response of San Joaquin Valley farmers to water supply cutbacks during the 1986–1992 drought, and two models that are commonly used to predict the effects of water supply reductions.

The work was sponsored by the U.S. Environmental Protection Agency (EPA) and RAND. The research was lead by Lloyd Dixon at RAND and Larry Dale, at the Law & Economic Consulting Group, Inc., Emeryville, California.

This report should be of interest to researchers and the policymaking community involved in drought issues. Related RAND research on water policy issues that might be of interest to the readers of this report include the following:

- *Drought Management Policies and Economic Effects in Urban Areas of California, 1987–1992,* Lloyd S. Dixon, Nancy Y. Moore, and Ellen M. Pint, MR-813-CUWA/CDWR/NSF, 1996.

- *Groundwater Recharge with Reclaimed Water: An Epidemiologic Assessment in Los Angeles County, 1987–1992,* Elizabeth M. Sloss, Sandy A. Geschwind, Daniel McCaffrey, and Beate R. Ritz, MR-679-WRDSC, 1996.

- *California's 1991 Drought Water Bank: Economic Impacts in the Selling Regions,* Lloyd S. Dixon, Nancy Y. Moore, and Susan W. Schechter, MR-301-CDWR/RC, 1993.

- *Assessment of the Economic Impacts of California's Drought on Urban Areas: A Research Agenda,* Nancy Y. Moore, Ellen M. Pint, and Lloyd S. Dixon, MR-251-CUWA/RC, 1993.

These reports are all available at nominal cost and can be obtained from RAND.

CONTENTS

FIGURES

TABLES

SUMMARY

In December 1994, the federal government promulgated new water quality requirements for the San Francisco Bay and the Sacramento and San Joaquin river systems and set critical habitat requirements for the delta smelt. Attaining these goals will require increased fresh water flow through the San Francisco Bay/Delta, which will, in turn, reduce the amount of water available to agricultural and urban users. California's State Water Resources Control Board (SWRCB) is currently conducting water rights hearings to determine how these reductions will be split between urban and agricultural users.

Of key concern to policymakers, farmers, and other stakeholders is how agriculture will respond to water supply cutbacks. Analyses of the impact of water supply reductions on agriculture usually rely on economic models of water use, but it is hard to verify the accuracy of these models. This report attempts to provide some insight into how water cutbacks might affect agriculture in the San Joaquin Valley. To this end, we examine

- what effects might be expected from economic theory
- previous empirical research on the effects of water supply reductions
- the effects of reduced water supplies on San Joaquin Valley agriculture during the 1986–1992 drought
- predictions of two models commonly used to estimate the effects of water supply cutbacks.

FARMER RESPONSE TO WATER SUPPLY REDUCTIONS: THEORY

There are a number of responses that farmers can make to reductions in water supplies. They may change the

- types of crops planted (the crop mix)
- amount of acreage planted (crop fallowing)
- irrigation technology (whether row, flood, sprinkler, or drip irrigation)
- irrigation management practices (the way farmers operate a particular technology)
- amount of water applied to crops (which may affect crop yield and the amount of salt buildup in the soil).

The type and degree of response will depend on whether farmers view the cutbacks as temporary or permanent, the amount of time they have to adjust, and the alternative sources of water available. Permanent water cutbacks may cause farmers to change their desired crop mix, for example; while temporary cutbacks may cause fallowing of certain crops but leave the desired crop mix unchanged. Adjustment time plays an important role in characterizing farmer response. It may be too expensive and time consuming to quickly acquire and learn to use the equipment to grow new crops; for example; and changes in desired crop mix may occur only gradually. Farmers may attempt to offset reduction in surface water deliveries with other sources of water. Most, but not all, farmers in the San Joaquin Valley have access to groundwater and increase groundwater pumping in response to surface water cutbacks. The effect of surface water cutbacks will thus depend on whether groundwater is available, whether or not it was the marginal source of water prior to the surface water cutback, and how the cost of pumping groundwater changes over time (for example because of declines in the water table due to increased pumping).

Table S.1 summarizes the type of changes we might expect in response to permanent and temporary water cutbacks, and the likely rate of adjustment. Both the degree and rate of adjustment will depend on the change in the marginal cost of water (or shadow price when supplies are limited).

Table S.1

Theoretical Farmer Response to Water-Supply Reductions

Response	Long-Run Response to Permanent Water Cutbacks	Response to Temporary Cutbacks	Adjustment Time	Demand for Farm Inputs
Desired crop mix	Possible shift from field crops to fruits, nuts, and vegetables	Little response	Gradual	May increase
Crop fallowing	Likely in the short run, but long-run response hard to predict	Likely, but unlikely for fruit, nuts, and vegetables	Rapid	Usually will decrease
Irrigation technology	Shift from flood and furrow irrigation to sprinkler and drip	Little response likely	Gradual	Skilled labor may increase, total labor may decrease
Irrigation management	More careful operation of given irrigation technology	More careful operation of given irrigation technology	Rapid	May increase somewhat
Deficit irrigation	Not a likely long-run strategy	Possible response	Rapid	Likely to have little effect on non-water inputs

Farmers and landowners will certainly be worse off with surface water reductions, but the effect of water cutbacks on other participants in the agricultural economy is not clear-cut. As shown in the last column of Table S.1, a cutback-induced switch from field crops to fruits, nuts, and vegetables would likely increase the demand for farm inputs **and** farm labor. The demand for downstream food handling and processing services may also expand. Shifts to new irrigation technologies may provide a significant stimulus to local irrigation businesses but may depress the demand for low-skill labor. Crop fallowing, however, would decrease the demand for farm inputs, labor, and the services of handlers and processors, although labor market effects may be dampened if farmers decide to hold on to their labor force during temporary cutbacks.

EXISTING EMPIRICAL STUDIES OF FARMER RESPONSE TO WATER CUTBACKS

There have been a number of empirical studies that shed light on how farmers respond to water cutbacks. The second column of Table S.2 summarizes findings from studies on the long-run response to permanent water supply cutbacks. There is some evidence that once cutbacks become sufficiently large, farmers move away from crops with high water, but low capital and labor requirements (particularly alfalfa). Consistent with expectations, there is also evidence that they will adopt irrigation technologies with higher irrigation efficiencies.

Table S.2

Summary of Empirical Evidence on Farmer Response to Water-Supply Reductions

Response	Long-Run Response to Permanent Reductions	Response During 1986–1992 Drought
Use of alternative water supplies	not addressed	Increased groundwater pumping; purchases of supplemental surface water
Desired crop mix	Less alfalfa, more wheat and barley; less field crops, more vegetables	Some weak evidence on shifts from field crops to vegetables
Crop fallowing	not addressed	Substantial fallowing of field crops
Irrigation technology	Sprinkler and drip replaces flood and furrow	Conflicting opinions on adoption of sprinkler and drip
Irrigation management	not addressed	Widespread improvements
Deficit irrigation	not addressed	Conflicting opinions
Employment and use of inputs	not addressed	Minimal impact

The last column of Table S.2 characterizes the findings from empirical studies of farmer response to water supply reductions in the San Joaquin Valley during the 1986–1992 drought. The studies suggest that crop fallowing and improvements in irrigation management were widespread and that groundwater pumping increased significantly to offset reductions in surface water supplies. There is disagreement on the extent of crop shifting and changes in irrigation technology, and not much evidence of substantial employment effects.

FARMER RESPONSE TO WATER SUPPLY REDUCTIONS: NEW EMPIRICAL EVIDENCE

To provide additional insight into the response of farmers to water supply reductions, we compare changes in agricultural activity during the 1986–1992 drought in two counties in the southern San Joaquin Valley (Fresno and Kern), where water use declined substantially, with three counties in the northern San Joaquin Valley (Merced, San Joaquin, and Stanislaus), where water use changed little. Comparing the two sets of counties helps isolate changes in agricultural activity caused by factors other than changes in water supply; the county is the unit of analysis because the county is the most desegregated level at which many economic and agricultural data are available.

We cannot be sure that the three northern counties are indeed good controls for Fresno and Kern counties, and our analysis is hampered by a relatively small number of counties and few years over which we have data. However, if carefully interpreted, we think the data informative.

Changes In Water Use During the Drought

There is a great deal of uncertainty about how agricultural water use actually changed during the drought. Surface-water diversions from the Central Valley Project and State Water Project are generally well monitored and the data readily available. Diversions by users with appropriative water rights on local rivers are also generally monitored, but the data are usually only available from local water districts. Riparian and groundwater use, in contrast, are very poorly monitored. The lack of data on groundwater use on a regional basis is particularly unfortunate in examining the impact of the drought because farmers in many parts of the San Joaquin valley are thought to increase groundwater pumping when surface water supplies decline.

In the body of the report, we consider two different measures of groundwater use. Table S.3 summarizes our best estimate of how surface, ground, and total water use changed in Fresno and Kern counties (the *impact* counties) and the control counties during the drought. The drought years are grouped based on patterns of surface water use observed in the impact counties.

There were large reductions in surface water deliveries in Fresno and Kern counties during the drought. These reductions were, to a substantial extent, offset by pumping of higher-priced groundwater. Exactly how much the reductions were offset is uncertain, but our best guess is that they were completely offset during the first years of the drought and only partially so in later years of the drought. We estimate that total water use fell 15 percent between average use in 1987–1989 and 1991, and 7 percent between average use in 1987–1989 and 1992.

Table S.3

Changes in Water Use in the Impact and Control Counties Between 1985 and 1992
(average percentage change)

	Surface Water	Ground-Water	Total Water
1985–1986 to 1987–1989[a]			
Impact	−18	32	0
Control	−5	8	−1
1987–1989 to 1991			
Impact	−70	47	−15
Control	−10	20	2
1987–1989 to 1992	−53	45	−7
Impact	−12	30	5
Control			

[a]Hyphenated years refer to average annual water use during those years.

The absolute, as well as percentage, declines in the latter part of the drought were substantial. Between 1987–1989 and 1991, surface water use in Fresno and Kern counties fell approximately 2.5 million acre-feet, and total water use fell approximately 975,000 acre-feet.[1] The substantial fall in water use in between 1987–1989 and 1991 and between 1987–1989 and 1992 implies that farmers either cut back use because they switched to a higher cost marginal water supply, or that they had limited or no access to groundwater, or both.

In contrast, the data suggest that the much more moderate cutbacks in surface water supplies in the control counties were completely offset by increased groundwater pumping. The data available also suggest that groundwater costs in the control counties are similar to surface water prices, implying that surface water reductions in the control counties had little effect on agricultural activity in the control counties during the drought.

Presumably, water cutbacks during drought would usually be interpreted by farmers as temporary. However, there is some evidence that farmers viewed at least part of the cutbacks during the 1987–1992 drought as permanent: Land values fell more in the impact counties than in the control counties. Some of the surface water cutbacks during the drought were due to more-stringent environmental regulations, which may have been viewed as permanent, and farmers and investors also may have viewed at least part of the water cutbacks as indicative of the new regulations to come.

Changes in Agricultural Activity

Table S.4 summarizes findings from our analysis of the effect of water supply reductions on San Joaquin Valley agriculture during the 1986–1992 drought. Data on cropping pattern are consistent with a steady shift from field crops to vegetables and from

---

[1]In comparison, a recent study predicts that the new environmental regulations will cause surface water deliveries in the San Joaquin Valley to fall 364,000 acre-feet in an average water year and 815,000 in a critically dry year.

low- to high-value field crops induced by water supply cutbacks.[2] Unfortunately, however, these shifts were not well correlated with changes in surface water deliveries, and we cannot be certain that they are due to water supply changes. It may be that these shifts represent a gradual response to perceived permanent surface water reductions; they may also be due to factors other than water supply.

Crop fallowing is widely expected in response to temporary water cutbacks and also expected as a short-run response to permanent water cutbacks, and there was, without a doubt, substantial field crop fallowing during the drought. It appears that farmers fallowed both low- and high-value field crops.

Changes in irrigation technology and irrigation management are likely in the response to permanent water cutbacks, and changes in irrigation management and deficit irrigation are likely in response to temporary cutbacks. As summarized in Table S.4, we found no evidence that farmers stressed their crops enough to reduce yields during the drought. Surprisingly, we also found no evidence in lasting improvement in irrigation efficiency (which could be because of changes in irrigation technology or management), but the data we had available for the analysis were weak.

Data on farmer profits and land values suggest that farmers did suffer losses due to the drought. However, the effect on agricultural employment, both on and off the farm, is less clear. We found no convincing evidence that the water supply reductions caused a fall in agricultural employment, although under some assumptions it is possible to conclude that on-farm crop production did fall. Employment effects due to temporary water cutbacks may be different from permanent cutbacks, but our analysis does leave open the possibility that the effect of permanent water supply cutbacks—in the range of those observed during the drought—may not be great.

MODEL PREDICTIONS OF THE IMPACT OF WATER SUPPLY CUTBACKS

We examine two economic models that are commonly used to predict the impact of water supply cutbacks in California: the Rationing Model, initially proposed by researchers at the University of California at Berkeley, and the Central Valley Production Model (CVPM), developed by researchers at U.C. Davis and California's Department of Water Resources. Even though these models are widely used by regulatory agencies, surprisingly little is known about the realism of their assumptions and the accuracy of their predictions.

The rationing model is transparent and simple to run. It assumes that farmers respond to water cutbacks by fallowing the crops with the lowest net or gross revenue per acre-foot of water applied until the amount of water saved equals the cutback. The main drawbacks of this model, however, are that there is no firm theoretical basis for such an assumption and that the change in groundwater pumping must be specified.

_____

[2]High-value field crops are defined as those with gross revenue greater than $500 per acre.

Table S.4

**Summary of Effects of Water Supply Reductions in Fresno and Kern Counties
Between 1986 and 1992**

Measure of Activity	Summary
Crop pattern:	
Acreage harvested	15 to 20 percent of field crops fallowed, both low- and high-value field crops
Crop mix	Weak evidence of shift from field crops to vegetables and from low- to high-value field crops
Irrigation practices:	
Crop yield	No evidence of reduced yield
Irrigation management	Water use data consistent with improvement in irrigation management in 1991, but deficit irrigation also a possible explanation
Irrigation technology	No indirect evidence of changes in irrigation technology
Value of production:	
Crop production	Reduced field crop production reduces overall crop production value 6 to 7 percent from 1987–1989 to '91 or '92; half of this decline offset by increased vegetable production
Livestock/poultry production	Somewhat faster growth in the impact counties relative to controls
Employment:	
Crop production	No clear evidence of any effect, but 5 percent reduction in on-farm crop production possible
Livestock/poultry production	No negative effect
Countywide	No negative effect detectable
Farmer profit	Approximately a 4 percent decline
Land values	Fell approximately $125 per acre in impact counties relative to controls
Access to credit	No change in loan rates, reduced access to credit for those with uncertain water supplies

The CVPM assumes that farmers act to maximize farm profits subject to water (and other resource) constraints and market conditions. It is on solid theoretical ground, but it is very complex and requires a very large number of input parameters. Its predictions are also limited by a large number of individual constraints on the amount of various crops that can be grown in particular regions. It is difficult to know how appropriate these constraints are for different types of simulations and how they affect the results.

We ran the rationing model and CVPM to predict the impact of water supply reductions in Fresno and Kern counties between 1987–1989 and 1991. Neither predicted a shift to growing vegetables—a shift that may have been due to water supply reductions or to other factors. The CVPM more accurately predicted the reduction of field crop and total acreage than the rationing model. Likewise, the CVPM more accurately predicted the split between the fallowing of low- and high-value field crops. The rationing model predicted that fallowing will be almost exclusively restricted to low-value field crops. The CVPM, in contrast, predicted greater acreage reductions in high-value than low-value field crops, which was much closer to reality. As expected, because it concentrates fallowing on the low-value field crops, the rationing model predicted a lower reduction in gross crop revenue than the CVPM did. The observed decline in crop revenue lay between the two model predictions.

EVALUATION AND POLICY IMPLICATIONS

New regulations on water quality in the San Francisco Bay/Delta will likely mean permanent reductions in surface water deliveries in the San Joaquin Valley.[3] In the short-run, farmers may offset part or all of these declines with increases in groundwater pumping. Over time, however, groundwater levels will decline, increasing the cost of groundwater and reducing the amount of groundwater pumped. It thus seems likely that a large fraction, if not all, of the reduction in surface water supplies will ultimately be reflected in a reduction in total water use.

Our review of theory, past empirical work, the effect of the 1986–1992 drought, and models used to predict the effect of water supply reductions offers the following lessons on the impact of permanent water supply reductions on the San Joaquin Valley.

Poor Data on Water Use Makes Analysis Difficult

The lack of reliable data on groundwater pumping in the San Joaquin Valley makes it very difficult to determine how farmers respond to water supply cutbacks. The lack of good data makes it difficult to use past experience to predict how future regulatory water reductions will affect agriculture. Better measurement of actual water use is needed.

Improvements in Irrigation Efficiency in Response to Permanent Reductions Likely Over the Long Run

There is a strong theoretical argument and strong empirical evidence that farmers will improve irrigation efficiency in response to permanent water supply reductions. We do not have good data with which to examine the change in irrigation technology and management during the drought, but other researchers have concluded that there were widespread improvements in irrigation management—although whether there was much shift to sprinkler and drip is under dispute. One expects to see such changes occur only gradually and in response to permanent, not temporary, cutbacks.

---

[3]There will still be variation across years, but the mean around which annual deliveries fluctuate will be lower.

The Effect of Water Supply Reductions on Crop Mix Remains Uncertain

Theory suggests that an increase in the shadow price of water may induce farmers to shift to crops that need less water or that have higher labor or capital intensities. There has been some empirical support for such changes in the past, but the evidence is limited. There was an increase in vegetable acreage during the drought in the impact counties, but uncertainty remains about whether this was due the water supply cutbacks or to other factors unrelated to water supply. Similar uncertainties remain about the shift from low- to high-value field crops. Any such shifts would be limited by the effect of increased production of, say, vegetables on vegetables prices.

Employment Effects of Long-Term Supply Reductions Ambiguous

The effects on long-term water supply reductions on agricultural employment are ambiguous in theory. Shifts to more labor-intensive crops or irrigation systems may offset any change, say, in the amount of acreage farmed.

There was no clear reduction in agricultural employment during the drought. Not only was there no discernible effect on overall employment in Fresno and Kern counties, but there was not even strong evidence that on-farm crop production employment fell. It may be that our employment data do not capture significant effects among seasonal or undocumented workers, that farmers held on to employees during what they perceived to be, in part, temporary water cutbacks in order to maintain their labor force, or that shifts to vegetables unrelated to the water supply cutbacks masked employment reductions during the drought. More work and better data are needed to sort out these possibilities, but for the time being, the impact of even substantial water cutbacks on agricultural employment remains uncertain.

Farmers and Landowners Are Adversely Affected by Water Reductions

Water supply reductions, whether they are temporary or permanent, will negatively affect farmers. Reductions represent constraints on farm operations and farmers can be no better off with them than without them. Agricultural land value represent the expected long-run profitability of farming and thus will also likely fall in response to water supply cutbacks. Data suggest that San Joaquin farmers and landlords were adversely affected during the drought: Farmer profits fell as did relative land prices in the counties that faced the largest water supply cutbacks.

Rationing Model Is a Limited Tool for Policy Analysis, Central Valley Production Model Needs to Be Better Understood

The rationing model provides first-order approximations for at least some farmer responses to the type of water supply reductions that occurred during the 1986–1992 drought. However, it does not do well at a more detailed level, and its central underlying assumption—that farmers fallow crops with the lowest net revenue per acre-foot of water applied—appears incorrect. Its inability to predict changes in groundwater pumping is also a major drawback.

The rationing model's simplicity and ease of use is tempered by this limitation and the inaccuracy of many of its predictions.

The CVPM appears to be a better choice for policy analysis, but the appropriateness of its many constraints needs to be better understood. Its complexity and intense data requirements temper its apparent accuracy in predicting responses to the 1986–1992 drought.

Ongoing Evaluation of Farmer Response to Water Supply Reductions Needed

Many uncertainties remain about how farmers will respond to permanent water supply cutbacks, and ongoing study of farmer response to the water supply reductions is needed. The empirical analysis in this report suffered from a limited number of counties and a limited number of years of data. Data on more counties and farther back in time would certainly help to resolve some of the uncertainties we faced in interpreting the numbers and should be collected. A more promising approach may be to analyze response to water supply cutbacks at the farm level, and it would also be productive to use cross sectional data to further examine the effect of changes in water price and water availability on farm practices. It is difficult and expensive to collect this micro-level data, but the resulting findings could be much more definitive than analyses using aggregate data.

Better information on the effects of water supply reductions on agriculture will allow policymakers to revisit decisions to reallocate water from agriculture to the environment with more accurate information on the costs and benefits.

ACKNOWLEDGMENTS

In the course of this project, we interviewed and requested information from many people involved in all aspects of the agricultural economy and in the water industry. This report would not have been possible without their cooperation and assistance. We cannot name them all, in part to protect confidentiality; however, we want to explicitly thank those who made special efforts.

We thank Ken Budman and Erlinda Cruz of the California Employment Development Department; Steve Hatchett of CH2M-Hill; Roger Putty of Montgomery Watson; Ray Hart, Ray Hoagland, Nasser Batani, Farhad Farnum, and Tom Harding of the California Department of Water Resources; David Shaad and Bob Briney at the Agricultural Stabilization and Conservation Service; Christopher Doherty of Western Farm Credit Bank; and Steve Kritcher of Mutual of New York.

We are also grateful to Gary Zimmerman of the Federal Reserve Bank of San Francisco; John Mamer and Curtis Lynn, both formerly with U.C. Cooperative Extension; Marian Porter at the California Department of Social Services; Don Villarejo at the Rural Studies Institute; and Linda Wear of Northwest Economic Associates.

Adele Palmer at RAND and David Zilberman at U.C. Berkeley formally reviewed the report, and we appreciate the care, speed, and insight with which they completed their task. Michael Hanemann at U.C. Berkeley, Terry Erlewine at the State Water Contractors, and Roger Putty also provided valuable comments on earlier drafts.

Several others at RAND made important contributions to the project. Steve Garber helped sort through some of the economic modeling issues; Roberta Shanman tracked down several data sources; Christina Pitcher edited the final report; and Pat Williams entered much of the data and assembled and corrected the various drafts. RAND's Institute for Civil Justice provided the additional funding needed to publish this report.

Finally, we would like to give special thanks to Palma Risler at the U. S. Environmental Protection Agency, Region 9. As the project officer, Palma helped develop the overall goals of the project, provided important institutional and policy background, and put us in contact with the right people.

1. INTRODUCTION AND BACKGROUND

In December 1994, the federal government promulgated new water quality requirements for San Francisco Bay and the Sacramento and San Joaquin river systems under the federal Clean Water Act and set critical habitat requirements for the delta smelt under the Endangered Species Act. Attaining these requirements necessitates increased fresh water flow through the San Francisco Bay/Delta and will reduce the amount of water available to California's agricultural and urban users. California's State Water Resources Control Board (SWRCB) is currently conducting water rights hearings to determine how these reductions will be split between urban and agricultural users and among agricultural users.

The impact of water supply reductions on agriculture is of key concern to policymakers, farmers, and other stakeholders. Analyses of the agricultural impact of water supply reductions usually rely on economic models of water use. It is hard to verify the accuracy of these models, however. This report attempts to provide some insight into how water cutbacks might affect agriculture in the San Joaquin Valley.[4] To this end we examine
- what effects might be expected from economic theory
- previous empirical research on the effects of water supply reductions
- the effects of reduced water supplies in the San Joaquin Valley during the 1986–1992 drought
- predictions of two models commonly used to estimate the effects of water supply cutbacks.

In the remainder of this section, we first provide background on the water quality regulations that affect the San Francisco Bay/Delta. We then discuss how much these regulations might affect agricultural water supplies and how these reductions compare to cutbacks during the 1986–1992 drought. We conclude by outlining the remainder of the report.

WATER QUALITY REGULATIONS FOR THE SAN FRANCISCO BAY/DELTA
The San Francisco Bay/Delta is the hub of California's water system. Water flows into the delta from the Sacramento and San Joaquin river basins. Between 1980 and 1992, approximately 21 million acre-feet per year flowed out to the ocean on average, and approximately 5 million acre-feet was exported south to the San Joaquin Valley and Southern California (California Department of Water Resources—CDWR, 1994, p. 250). Annual outflows and exports vary significantly depending on the amount of rainfall. The federal Central Valley Project (CVP) and the State Water Project (SWP) are the principal water exporters from the delta.

The primary water quality regulations for San Francisco Bay are

---

[4]The San Joaquin Valley covers the southern half of California's Central Valley.

- Decision 1485 (D-1485) issued by California's State Water Resources Control Board in 1978
- the biological opinion for winter-run Chinook salmon issued by the National Marine Fisheries Service (NMFS) in 1991
- the Central Valley Project Improvement Act of 1992 (CVPIA)
- water quality standards and the designation as a critical habitat for the delta smelt issued by the U.S. Environmental Protection Agency (EPA) and the U.S. Fish and Wildlife Service (USFWS) in December 1994.

D-1485 required that the CVP and SWP make operational adjustments to keep delta water quality and fresh-water outflow within specified limits. Fish and wildlife resources continued to decline after 1978, however, and EPA, NMFS, and the USFWS responded with more-stringent environmental regulations. In the case of the NMFS, decline of winter-run Chinook salmon, which is listed as a threatened species, prompted action under the Endangered Species Act. EPA and USFWS responded to the continued decline of a wide range of species with action under the Clean Water Act and Endangered Species Acts. With the passage of the CVPIA, Congress also set aside 800,000 acre-feet of the approximately 8 million acre-feet diverted by the CVP annually for environmental uses.

IMPACT OF THESE REGULATIONS ON DIVERSIONS TO SAN JOAQUIN VALLEY AGRICULTURE

What these regulations will mean for water supplies to San Joaquin Valley agriculture depends on the still ongoing water rights process. Reductions to the San Joaquin Valley are expected to be substantial. An analysis done in 1994 by EPA suggested that deliveries to San Joaquin Valley agriculture would fall 600,000 acre-feet on average and 1.3 million acre-feet in critically dry years relative to the requirements under D-1485 (U.S. EPA, 1994, Table 5-2).[5,6] More-recent analyses by the SWRCB project that deliveries to San Joaquin Valley agriculture will drop 367,000 acre-feet in an average water year and 815,000 acre-feet in a critically dry year (California SWRCB, 1997, p. V-3). These reductions will mostly likely be concentrated in certain portions of the San Joaquin Valley—those with the most junior water rights—and will vary considerably depending on the type of water year.

Farmers in the southern San Joaquin Valley faced severe reduction in surface water supplies during the 1986–1992 drought. As will be discussed in more detail below, surface water supplies fell over 3 million acre-feet (nearly 75 percent) in Fresno and Kern Counties between 1985 and 1991, the worst year of the drought.[7] Farmers partially offset this decline

---

[5]We adjusted the EPA estimates to include the NMFS requirements. We assumed that the entire increment in water cutbacks due to the NMFS requirements is borne by agriculture. EPA did not assume that any of the 800,000 acre-feet set aside by the CVPIA would be available to offset the reductions to agriculture.

[6]Critically dry years are roughly the 10 percent of years with the lowest water exports.

[7]Fresno and Kern Counties are the two largest counties in the southern San Joaquin Valley.

by increasing groundwater use, but overall water use still fell approximately 1.2 million acre-feet (18 percent) in the two counties between 1985 and 1991.

The large reduction in water use during the drought provides an opportunity to examine how farmers respond to water supply cutbacks. As will be discussed in Section 2, caution must be taken in inferring the effects of regulatory cutbacks from the effects of drought cutbacks. Drought is a temporary phenomenon whereas regulatory cutbacks are likely to be permanent. Nevertheless, the response of farmers during the drought may provide some lessons on how they might respond to permanent water cutbacks.

OUTLINE OF REPORT

In Section 2, we first discuss the types of response we might expect to water supply cutbacks and fill in the theory with description of the crop production opportunities available to farmers in the San Joaquin Valley. We then review the existing empirical literature on farmer response to water supply cutbacks, both in the San Joaquin Valley during the drought and in other settings. In Section 3, we examine new data on the impact of water supply reductions in the San Joaquin Valley during the drought by comparing changes in agricultural activity in counties that saw large declines in water use (Fresno and Kern) with counties where there was little change. Our analysis is based on countywide data collected from county agricultural commissioners, California's Employment Development Department, and the U.S. Bureau of Reclamation. In Section 4, we describe two models that have been commonly used to simulate farmer response to regulatory reductions and use them to project the impact of the water supply reductions observed during the drought. We then compare the projections to the actual changes in agricultural activity observed during the drought. The report concludes by summarizing the main lessons learned from the analysis.

2. FARMER RESPONSE TO WATER SUPPLY REDUCTIONS: THEORY AND EXISTING EMPIRICAL EVIDENCE

The impact of water supply reductions on the farm economy is the result of many decisions made by many different actors. Regional and local water districts decide how surface water supply reductions are distributed among farmers. Farmers decide how to change their farming operations in response to these reductions. Farmer decisions ultimately determine the resulting change in acreage planted, agricultural employment, and farm revenue. In this section, we first discuss the economic theory of farmer decisionmaking and then discuss what this theory suggests for how farmers may respond to water cutbacks and how these responses may affect the overall farm economy. We then review existing empirical studies on farmer response to water cutbacks, both over the long run and between 1987 and 1992 in the San Joaquin Valley.

FARMER RESPONSE TO WATER SUPPLY REDUCTIONS: THEORY

Profit Maximization

Economic analysis of farmers response to water supply reductions starts with the assumption that farmers attempt to maximize profit. In deciding what crops to grow; what irrigation technologies to use; and what combination of inputs such as water, labor, fertilizer, and pesticides to use; economists usually assume that farmers attempt to maximize the profitability of their farm operation. There are many subtleties to this profit-maximizing behavior. Two subtleties that are important in predicting farmer response to water cutbacks are profit maximization over time and the role of risk. We briefly discuss each in turn.

To varying degrees, farmers will maximize the profits they earn over time, as opposed to during just one growing season. For example, they may plant crops, such as tree and vine crops that take several years to mature but are productive over many years. To maintain the long-term productivity of the soil, farmers usually must rotate crops. This means that a farmer may grow some crops that, when considered in isolation, are only marginally profitable, or even unprofitable. Thus when maximizing profits over time, a farmer must consider the profitability of whole cropping rotations over time rather than the profitability of individual crops in a given growing season.

Economists assume that to varying degrees farmers also trade off profitability and risk when making decisions—preferring less uncertainty in ultimate profits for any given level of expected profits.[8] Different crops are associated with different levels of risk. It is commonly thought that both the yield risk and the market risk are high for vegetables.

---

[8]In economics terminology, farmers are *risk averse*: If two farming operations have the same expected profits, farmers prefer the one that has lower variance of profitability around the mean.

Yield risk refers to the variation in crop yield caused by weather, disease, or other factors beyond the farmer's control. Market risk refers to variation in the unit selling price for the crop. Government crop commodity programs can help a farmer reduce the variability in overall profit. Commodity programs for such crops as wheat, corn, and cotton remove both the yield risk and the market risk by ensuring a farmer a target price for a guaranteed quantity on each acre enrolled in the program.[9] Thus, farmers may grow field crops that appear to have a lower expected return than other crops.

The Permanence of Water Cutbacks and Adjustment Time

Farmers can make a variety of changes in their farm operations in response to surface water cutbacks. Farmers can obtain water from other sources if available, fallow land, shift crops, shift irrigation technologies, improve irrigation management, or perhaps go out of business altogether.

A key to understanding farmer response to water supply reductions is whether farmers view the cutbacks as temporary or permanent and the amount of time over which farmers can adjust.[10] Figure 2.1 schematically illustrates how the permanence of the water cutback and adjustment time interact.

Permanent surface water cutbacks, such as those due to regulatory cutbacks, may cause farmers to change the desired character of their farming operations.[11] They may

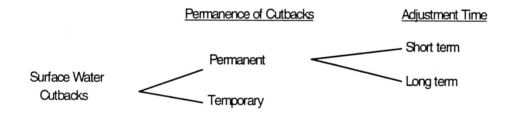

Figure 2.1—Interaction of the Permanence of Water Supply Cutbacks and Adjustment Time

---

[9]In return, a farmer must agree not to grow the crop on a certain fraction (often 10 percent) of the acreage enrolled in the program.

[10]Surface water deliveries vary from year to year because of weather fluctuations and can be thought of as a random variable with a mean and a variance. By temporary cutbacks we mean cutbacks due to normal variation in surface water deliveries that leave the underlying mean and variance of surface water deliveries unchanged. By permanent cutbacks we mean cutbacks that lower mean surface deliveries. Such cutbacks may also change the variance.

[11]By *desired* character of their farming operation, we mean the farmer's target cropping pattern, irrigation technology, labor use, etc. The farmer may be forced to deviate from the desired character of the farming operation by any number of external factors.

switch crops, shift irrigation technologies, the amount of land cultivated, or perhaps the mix of inputs for particular crops. It may make sense economically to make many of the changes only gradually. It may be too expensive and time consuming to learn how to grow many new crops quickly. It may be difficult to secure the credit to finance a wholesale transformation of the farm operation. Thus, farmers may be able to fully adjust to the permanent cutbacks only over a long period of time, and minor changes will be observed only in the short run.[12]

Farmers presumably would not want to make any adjustments to the underlying character of their farm operation in response to temporary surface water cutbacks that are expected as part of the normal variation in rainfall over time. This does not mean, however, that they will not need to make dramatic changes, albeit temporarily. Examples include land fallowing, which may have a dramatic effect on farm income and employment. Droughts result primarily in temporary cutbacks, but to the extent that droughts provide evidence that past expectations of rainfall were too optimistic, they may signal permanent water cutbacks.

Farmer response to temporary cutbacks may resemble the short-run response to permanent water supply cutbacks in many ways. Land fallowing is a likely example. Conversely, the two types of adjustments may differ importantly. For example, permanent water changes may cause a farmer to pull out an elderly fruit orchard right away because the orchard may no longer be profitable in the long run, whereas, during a temporary cutback, the farmer may want to apply a minimum amount of water necessary to keep the orchard healthy.

Alternative Water Supplies

Farmers may attempt to offset reductions in surface water deliveries with other sources of water. We first discuss increased use of groundwater and then discuss purchases of surface water from other sources.

Increased Groundwater Use. Many farmers in the San Joaquin Valley, but by no means all, have access to both surface water and groundwater. Farmers with access to groundwater can offset surface water reductions with increased groundwater pumping. In some cases, they may be able to completely offset the reductions. In other cases, pump capacity may be insufficient to offset reductions in the short run, but pump capacity may be augmented in the long run.

The impact of reduced surface water deliveries on farms with access to groundwater will depend in part on whether groundwater was the marginal water source prior to the cutback.[13] Figure 2.2a illustrates such a situation. Graphed is the demand curve for water

---

[12]The short run is the period over which only minor changes in farm- or crop-specific equipment and knowledge can be made. In the long run, farmers are able to learn how to grow different crops and use different equipment.

[13]Groundwater is usually more expensive than surface water on a given farm in the San Joaquin Valley with access to both, and thus groundwater is the marginal water supply (e.g., see

for an individual farmer during a single growing season. The farmer uses q_1 units of surface water before the reductions and q_2 after, but total water use remains unchanged (q_T). (P_{sw} denotes the price of surface water and P_{gw} denotes the price of groundwater.) We thus might expect to see little change in farm operation of such farmers.[14] The farmers themselves, however, would see profits drop because of the increased total water bill. Figure 2.2a assumes that pumping cost does not increase with increased pumping; however, increased pumping will likely cause the depth-to-water, and thus pumping costs, to increase over time.[15] The surface water cutback will therefore likely affect farm operations over time.

Now consider farmers who have access to higher-cost groundwater but have adequate surface water before the cutback (see Figure 2.2b). Surface water reductions would then cause the cost of the last unit of water applied to increase and create a price incentive to reduce water from q_{T1} to q_{T2}.[16] We will discuss shortly the ways that farmers may reduce water use.

Northwest Economic Associates, 1992, p. 11). Data on average costs of surface water and groundwater in the San Joaquin Valley are presented below, in Table 3.1.

[14]In many cases, groundwater is of lower quality (higher salinity and/or lower temperature) than surface water, and increased groundwater use may reduce crop yields.

[15]Depth-to-water usually increases as pumping continues over the course of a single growing season. However, when groundwater levels are in equilibrium, water levels will recover by the beginning of the next growing season. Figure 2.2 does not consider increases in depth-to-water during the growing season. Significant changes in depth-to-water during a growing season may affect farmer decisions relative to the case where depth-to-water is constant.

[16]As will become apparent below, water costs can represent a significant proportion of overall growing costs for certain crops (e.g., field crops).

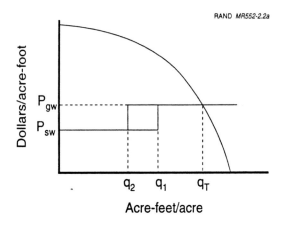

Figure 2.2a—Groundwater Is the Marginal
Water Source Both Before and After Surface
Water Reduction

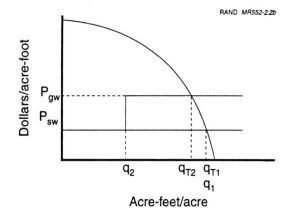

Figure 2.2b—Surface Water Is the
Marginal Water Source Before and
Groundwater Is Marginal Source
After Surface Water Reduction

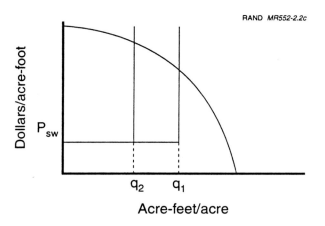

Figure 2.2c—No Groundwater Available

Figure 2.2c depicts a farmer who has no access to groundwater. The surface water
cutback thus forces the farmer to reduce usage from q_1 to q_2 with the consequent impact on
farm operations.

Purchases of Supplemental Surface Water. Whether or not farmers have access
to groundwater, they may attempt to purchase surface water to offset surface water
reductions. There is a well-established market for trading water among farmers within
water districts in the San Joaquin Valley and some opportunity to buy water from sources
outside the district. For example, farmers were able to buy water from the 1991 Drought
Water Bank set up by the State of California (see Dixon, Moore, and Schechter, 1993).

How much supplemental surface water farmers buy depends on its costs relative to
groundwater and the value of additional water in their farm operations. As in the case of
groundwater, if the higher-priced supplemental surface water became the marginal water

source, it would create a price incentive to reduce water use. Whether or not a farmer decides to purchase expensive supplemental water also depends to some extent on whether the farmer views the surface water cutback as temporary or permanent. A farmer may be willing to buy expensive water to keep perennial crops alive, for example, during a temporary cutback, but may not be willing if the cutback is permanent.

Summary. The impact of a surface water cutback depends importantly on how the marginal source of water changes and the price of the marginal source of water. If groundwater is the marginal source of water before the surface water reduction, there may not be much change in farm operations. If the marginal source of water is much more expensive, or if there is no marginal source of water, then there may be important changes in farm operations.

Responses to Surface Water Supply Reductions

We now discuss the principle types of adjustments farmers might make to cutbacks in surface water supplies and whether these adjustments are most likely in response to temporary or permanent cutbacks and in the short or long run. We also discuss how the adjustments might affect the farm economy and the use of farm labor in particular.

Shifts in Desired Crop Mix. A wide variety of crops can be, and are, grown in the San Joaquin Valley. This allows farmers to choose among widely differing combinations of inputs and crop outputs when responding to surface water cutbacks. For example, different crops require different amounts of water. Table 2.1 shows the evapotranspiration of applied water (ETAW) for crops that are commonly grown in the San Joaquin Valley.[17] ETAW varies significantly among crops—varying from 3.3 acre-feet per acre for irrigated pasture to 0.9 acre-feet per acre for wheat in the southern San Joaquin Valley. Crops also require different amounts of labor and other inputs. Table 2.2 shows that average production costs per acre, excluding water and land costs, vary significantly.[18]

---

[17]Crop evapotranspiration represents the biological water requirement of the crop. Evapotranspiration of applied water adjusts this requirement downward to account for the amount of water used by the plant that is provided by rainfall.

[18]These costs represent average or slightly better-than-average farming practices.

Table 2.1

Evapotranspiration of Applied Water for Crops Commonly Grown in the San Joaquin Valley (acre-feet per acre)

	Southern San Joaquin Valley[a]	Northern San Joaquin Valley[b]
Field crops		
Irrigated pasture	3.3	3.1
Alfalfa	3.1	3.1
Sugar beets	2.9	2.4
Cotton	2.5	2.4
Corn	2.0	1.9
Wheat	0.9	0.6
Fruits and nuts		
Almonds	2.3	2.0
Grapes	2.1	1.6
Citrus	1.9	1.6
Vegetables		
Tomatoes	2.1	1.9
Other truck crops	1.2	1.2

SOURCE: CDWR, 1994, Table 7-6.
[a]Primarily Fresno, Kern, Kings, and Tulare Counties.
[b]Primarily Madera, Merced, Stanislaus, and San Joaquin Counties.

Table 2.2

Average Production Costs for Crops Commonly Grown in the San Joaquin Valley Excluding Water and Land Costs (1994 dollars per acre)

Crop	Average Costs
Field crops	
Irrigated pasture	176
Wheat	214
Corn	297
Alfalfa	456
Sugar beets	533
Cotton	665
Fruits and nuts	
Almonds	930
Raisin grapes	1,075
Citrus	1,620
Vegetables	
Processing tomatoes	907
Potatoes	1,924
Melons	1,938
Fresh tomatoes	4,390

SOURCE: Crop production costs used in Bureau of Reclamation's version of CVPM.

Economic theory suggests that the desired mix of crops will change as water becomes scarce or the marginal cost of water increases.[19] How the optimal crop mix changes will depend on the characteristics of all the crops farmers can choose from—a detailed examination of which is beyond the scope of this study. We might expect, however, that farmers will shift to a crop mix that requires more labor and/or capital. As suggested by Table 2.2, we might thus expect to see shifts to fruits, nuts, and vegetables and away from field crops in response to permanent water reductions. Similarly, farmers might also tend to move away from crops that have the highest water requirement per acre. The high water requirements of several field crops (see Table 2.1) again suggest a movement away from field crops to fruits, nuts, and vegetables.

There are several factors that might limit any long-run change in crop mix from field crops to fruits, nuts, and vegetables. First, California, and the San Joaquin Valley in particular, accounts for a high proportion of U.S. and even world production of many fruits, nuts, and vegetables. Increases in San Joaquin Valley production of these crops may cause crop price declines that would diminish their profitability. Second, crop commodity programs for field crops provide a low-risk component to the farmer's portfolio. Farmers may thus want to keep a solid base of field crops in their crop mix. Third, some field crops may continue to be valuable as rotation crops.

As discussed above, shifts in crop mix will likely occur only gradually over time. It takes time for farmers to build up the knowledge a crop-specific equipment to grow new crops. Also, temporary cutbacks that are part of expected weather fluctuations are unlikely to cause changes in a farmer's desired crop mix, although as discussed below, the farmer may well take some land temporarily out of production.

The shift in crop mix may have significant impacts on the demand for farm inputs. A shift from field crops to vegetables, for example, would increase the demand for farm inputs (see Table 2.2) In particularly, a shift to fruits, nuts, and vegetables may increase the use of farm labor, but the magnitude and even sign of the effect will depend on the specific crop shifts. Table 2.3 shows that there is some overlap in the amount of labor required for the crops in the three crop categories, but by and large, labor requirements for fruit, nuts, and vegetables are higher than for field crops. Thus, particularly if fewer acres of irrigated pasture, wheat, corn, and alfalfa are grown, labor employment may actually rise in the new crop mix.

---

[19]Economic theory predicts that farmers will chose a cropping pattern such that the marginal revenue product of water for all crops is equal (taking into account risk and crop rotation benefits). In response to a water supply cutback, farmers will presumably choose crops whose marginal revenue product rises most rapidly as water use declines.

Table 2.3

Labor Required for Crops Commonly Grown in the San Joaquin Valley
(hours/acre)

Crop	Regular Employees	Temporary Employees	Total
Field crops			
Irrigated pasture	NA	NA	10
Wheat	NA	NA	10
Alfalfa	NA	NA	12
Corn	NA	NA	13
Cotton	24	3	27
Sugarbeets	31	24	55
Fruits and nuts			
Almonds	10	14	24
Oranges	23	80	103
Raisin grapes	22	82	104
Vegetables			
Carrots	17	0	17
Potatoes	18	6	23
Processing tomatoes	22	31	53
Snap Beans	14	74	88
Melons	13	77	90

NA: not available.

SOURCE: Mamer and Wilkie, 1990; except amounts for irrigated pasture, wheat, corn and alfalfa, which are derived from the University of California Agricultural Extension crop budgets.

Crop Fallowing. In response to both temporary and permanent water cutbacks, farmers may decide to take some crops out of production. In the case of an entire region or county, permanent water cutbacks may cause marginal farmland to be permanently retired from production. Temporary cutbacks may cause farmers to fallow farmland for a single growing season.[20]

Just as it is difficult to predict how farmers will change their crop mix, it is difficult to predict which crops, and how many acres, farmers will decide to take out of production in response to permanent water cutbacks. Again, it will depend on the characteristics of all crops farmers have to choose from. There do seem to be some sensible principles that should guide the fallowing of crops in response to temporary water cutbacks, however. First, it seems unlikely that farmers will fallow fruit and nut crops in response to temporary cutbacks. Fruit and nut trees and vines can produce yields for many years, and it may make economic sense for farmers to ensure their continued production. Second, farmers may give priority to crops for which they want to maintain marketing relationships. It can take years to build marketing arrangements for crops such as rice, and stable production may be a condition of the marketing arrangement. Third, many farmers have ownership

_____

[20]Farmers may fallow land in order to use the water allocated to that land on other planted acreage—a type of water farming.

interests in downstream processors and distributors for particular crops and may want to ensure enough production of those crops to keep their operations going.

Crop fallowing will likely reduce the demand for farm inputs. The reductions in non-labor inputs—such as seed, fertilizer, and pesticides—will depend in part on how the farmer had invested before the water cutback was announced.[21] Employment should, for the most part, fall; but during temporary cutbacks, some farmers may shift labor to other tasks around the farm to maintain their labor force for future years.

Improvements in Irrigation Efficiency. Farmers may respond to surface water cutbacks by attempting to increase their irrigation efficiency. They may try to do this either by changing irrigation technologies or by operating their existing irrigation systems more carefully.[22]

Farmers may be able to increase irrigation efficiency by shifting from flood and furrow irrigation to sprinkler and drip irrigation systems. Sprinkler and drip systems are more expensive to install, and perhaps maintain, than flood and furrow systems, but the water savings may prompt farmers to shift to sprinkler and drip when faced with water shortages or higher prices.[23] Currently flood and furrow irrigation are most common in the San Joaquin Valley (see Table 2.4), although the amount of sprinkler, microsprinkler, and drip have been growing in recent years (Zilberman et al., 1994).

Empirical studies of water use suggest that shifting to sprinkler and drip systems can indeed significantly reduce water use. Studies by Dixon (1988, Table 2.5) and CDWR (CDWR, 1986) suggest that farmers apply less water to the same crop with sprinkler and drip systems than with flood and furrow systems. Caswell and Zilberman (1985) report that, on average, the irrigation efficiency of traditional irrigation is about 60 percent, sprinkler irrigation about 85 percent, and drip irrigation 95 percent.

Table 2.4

Irrigation Technology in the San Joaquin Valley

Irrigation Technology	Percentage of 1991 Crop Acreage
Flood and furrow	71
Sprinkler	21
Drip	7
Subsurface	1
Total	100

SOURCE: CDWR, 1993, Chapter 7.

---

[21]Preliminary water deliveries estimates are usually not announced in the San Joaquin Valley until February or March with final deliveries announced in April. In many cases, farmers have already prepared the soil and even planted some crops by that time.

[22]Irrigation efficiency is the ratio of crop evapotranspiration to applied water.

[23]There are many advantages to some low-flow irrigation technologies that have nothing to do with water savings. For example, drip systems can apply fertilizer effectively and may increase yields because the plant does not need to develop as extensive a root system.

Table 2.5

Applied Water by Irrigation Technology in Kern County
(acre-feet per acre)

Crop	Flood and Furrow	Sprinkler	Drip
Field crops			
Cotton	3.4	3.1	
Alfalfa	4.4	4.0	
Grains	1.7	2.5	
Citrus fruits		4.3	2.0
Vegetables	3.0	2.8	

SOURCE: Dixon, 1988, p. 210.

Even if farmers do not change irrigation technologies, many management changes that improve irrigation efficiency may be possible. Management improvements may be relatively easy and inexpensive to make. Examples of management improvements for flood and furrow irrigation include laser leveling, tail water recovery, and careful monitoring of soil moisture. These management improvements may have a significant effect on water usage, and it may be that, in certain situations, irrigation efficiencies for flood and furrow may approach sprinkler and drip, *if properly managed* (CDWR, 1994, p. 166).[24]

A substantial shift to new irrigation technologies is likely only in response to permanent water cutbacks and will likely happen only gradually. In contrast, it may be possible to make many management improvements in the short run. As with changes in crop mix, there may be little change in irrigation technology if the cutback is viewed to be temporary, although relatively easy and inexpensive management improvement may well make sense.[25]

In the short run, or in response to temporary cutbacks, farmers may conserve water by deficit irrigating their fields. They may not apply enough water to flush out the salts that build up in the soil during the year.[26] This may not affect crop yield in the short run, but will likely have an effect on yield over the long run. Farmers may also reduce water application to the point where current yield is reduced. In deciding whether or not to "stress" their crops, farmers must trade off the water savings with the reduction in yield.

Shifts in irrigation technologies will likely change the demand for farm labor. There is not a great deal of quantitative information on how the labor requirements of sprinkler and drip differ from flood and furrow. However, it seems likely that a shift from flood and

---

[24]For example, Negri and Brooks (1990) point out that traditional gravity systems applied to land with high water-holding capacity due to high clay content and level slopes can achieve application efficiencies comparable to sprinkler irrigation.

[25]There are many advantages of drip and sprinkler irrigation independent of irrigation efficiency. Drip systems can be used to apply fertilizers with great precision. Drip systems may also help increase yield by requiring the plant to develop a smaller root system.

[26]Some estimate that approximately 0.5 acre-foot of water a year may be needed to adequately flush salts.

furrow would result in a reduction of labor hours overall but increase demand for certain skilled workers (Negri and Brooks, 1990, p. 214).

Summary of Likely Farmer Responses to Surface Water Reductions

We summarize our investigation of farmer response to surface water supply reductions first for permanent reductions and second for temporary reductions.

The impact of permanent water supply reductions will depend a great deal on farmer access to groundwater and access to supplemental water sources. The impact of surface water cutbacks on farm operations will be least severe if groundwater is already the marginal source of water and farmers can easily pump more. It will be more severe if higher cost groundwater becomes the marginal source, and the most severe in the extreme case that groundwater is so expensive that none is pumped or if there is no groundwater available.

The second column of Table 2.6 summarizes the type of changes we might expect in response to permanent surface water cutbacks, with the intensity depending on the change in the marginal cost of water. The desired crop mix may change over time in response to permanent water reductions. We did not come to any definite conclusions on how the desired crop mix might change, but thought it plausible that farmers would shift from field crops to perennial crops and vegetables—crops that are labor and capital intensive and use less water per acre. Farmers may also permanently retire marginal farm land from production in response to permanent water cutbacks—but the magnitude of the response is difficult to predict. Farmers are likely to use more sprinkler and drip irrigation and improve irrigation management in response to permanent water cutbacks. As summarized in the penultimate column of Table 2.6, it is likely that changes in cropping pattern and irrigation technology will occur only gradually over time, but some irrigation management changes might be made in the short run.

Table 2.6

Summary of Conceptual Discussion of Farmer Response to Surface Water Reductions

Response	Long-Run Response to Permanent Water Cutback	Response to Temporary Cutback	Adjustment Time	Demand for Farm Inputs
Desired crop mix	Possible shift from field crops to fruits, nuts, and vegetables	Little response	Gradual	May increase
Crop fallowing	Difficult to predict	Fallowing unlikely for fruit, nuts, and vegetables	Rapid	Usually will decrease
Irrigation technology	Shift from flood and furrow to sprinkler and drip	Little response likely	Gradual	Skilled labor may increase, total labor may decrease
Irrigation management	Increase irrigation efficiency	Increase irrigation efficiency	Rapid	May increase somewhat
Deficit irrigation	Not a likely long-run strategy	Possible response	Rapid	Likely to have little effect on non-water inputs

The availability of groundwater is also central to evaluating the impact of temporary water cutbacks. Temporary cutbacks may have little effect if groundwater pumping capacity and groundwater quality is adequate. As indicated in the third column of Table 2.6, we do not expect farmers to change their desired crop mix, but farmers may be able to make temporary adjustments. Fallowing will likely be restricted to field crops, as opposed to fruits, nuts, and vegetables. The desired mix of irrigation technologies should not change much in response to temporary water reductions because of the cost of installing new technologies and learning how to use them. However, farmers may improve irrigation management practices or deficit irrigate their crops.

Winners and Losers from Water Reductions. Farmers will certainly be worse off with surface water reductions. Reductions impose a constraint on their operations; farmers, therefore, can be no better off than before. How much worse off they will be depends on the extent to which they can adapt to the new circumstances.

If the water supply reductions are permanent, landowners may also be adversely affected by the drought. Economic theory suggests that agricultural land values are generally based on the capitalized value of the earnings potential of the land over a long period of time. Consequently, land values and land rents should decline with permanent water cutbacks because farm profitability will decrease.[27] In contrast, land values should be little affected by drought-induced cutbacks that are temporary in nature. Temporary water fluctuations that are part of expected weather variability should already be incorporated in land prices.

The effect of water cutbacks on other actors of the agricultural economy is not clear-cut. As shown in the last column of Table 2.6, a cutback-induced switch from field crops to fruits, nuts, and vegetables would likely increase the demand for farm inputs and farm labor. The demand for downstream food handling and processing services may also expand. Shifts to new irrigation technologies may provide a significant stimulus to local irrigation businesses but may depress the demand for less-skilled labor. However, crop fallowing would likely decrease the demand for farm inputs, labor, and the services of handlers and processors, although labor market impacts may be dampened if farmers decide to hold on to their labor force during temporary cutbacks.

EMPIRICAL STUDIES OF FARMER RESPONSE TO WATER CUTBACKS

There have been a number of empirical studies that shed light on how farmers respond to water cutbacks. We first examine studies that can be interpreted as capturing long-run response to permanent water cutbacks. We then examine studies on the impact of reductions in surface water supplies between 1987–1992 in the San Joaquin Valley. As we will argue, the impact of these reductions likely represents short-run response to reductions that had both temporary and permanent components.

Long-Run Response to Permanent Water Supply Reductions

There appear to be few studies that directly look at the long-run response to permanent cutbacks in water supply.[28] Rather than look for direct evidence, we examine indirect evidence of how farmers respond to water cutbacks over time—namely, we review studies on how farm operations vary with the price of water at a given point in time.[29] Water cutbacks are analogous to increases in water costs in the sense that water cutbacks should lead farmers to allocate water as though it had a higher price. In Figure 2.3, a cutback from q_1 to q_2 causes farmers to allocate water though it were priced at P_{sh}. This notional cost is called the shadow price of water.[30] In many agricultural settings water

---

[27]Land values are frequently used to secure property and equipment loans in the San Joaquin Valley. To the extent that water constraints cause some farmers to default on loans and lenders have to foreclose, lower land values may also leave lenders with an asset insufficient to cover the remaining debt.

[28]One reason may be that it is difficult to separate the effect of water cutbacks from the changes in many other factors that affect farm operations over time.

[29]These are referred to as cross-sectional studies of farm behavior (as opposed to time-series studies).

[30]Even though water price increases and surface water cutbacks may have similar impacts on the character of farm operations (such as the number of people employed), they certainly have different impacts on farm profitability—farmer water costs are much higher with price increases than cutbacks.

prices have been fairly stable over time, so farmers should have made long-run adjustments to the price levels.

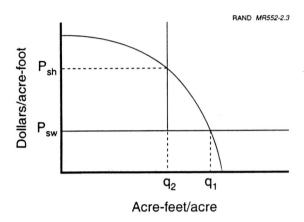

RAND MR552-2.3

Figure 2.3—Surface Water Reduction Raises the Shadow Price of Water

Before examining the effect of water price on cropping pattern and irrigation technology, several caveats are necessary on the inferences about farmer response to water reductions that can be drawn from studies of farmers facing different water prices at a given point in time. First, observed differences in cropping pattern between farmers facing different water prices may overstate the impact of a widespread water price increase or water cutback on all farmers. This is because widespread behavioral changes may affect the supplies of different farm commodities enough to change relative crop prices.[31]

Second, variation in farm operation across farmers may be due to factors other than water prices. For example, soil characteristics, input and output prices, and weather may cause the variation. Care must be taken to control for these confounding factors.

Third, farmer response to surface water cutbacks may not always be the same as the response to paying a higher price for water. For example, very large cutbacks may induce a farmer to hold some land for its water right but use the water elsewhere. Such a response would not be expected from a farmer who can obtain all the water desired at a certain high price. The range of prices and cutbacks over which the similarity of response to water cutbacks and price increases holds needs further study, but the similarity should hold for moderate cutbacks.

Cropping Pattern. In a study of four western multi-state regions, Moore, Gollehon, and Carey (1994) found that cropping pattern responded to water price. They restricted their attention to farmers that grew common field crops (such as alfalfa, corn, and cotton) and did not grow fruit, nut, or vegetable crops between 1984 and 1988. They found that overall farm water use was very insensitive to water price but that there were changes in

---

[31]A crop shift that may seem desirable from the point of view of an individual farmer may not make sense for an entire region if increased production causes the crop price to decline. Because the San Joaquin Valley produces a high percentage of the overall output of many crops in the United States and even the world, this effect may be significant there.

crop mix. In particular, they found that producers substituted away from alfalfa production in the face of higher water prices (p. 867).

In a 1988 survey of 35 farmers in Kern County, California, Dixon (1988) found a strong relationship between marginal water cost and cropping pattern.[32] As shown in Table 2.7, he found that when water cost rises above $45 per acre-foot there is a marked decline in field crops and a marked increase in vegetables. The percentage of fruit and nut observations remains fairly stable across the three price categories. The primary change among field crops is a drop in alfalfa, which appears to be partially offset by an increase in grains such as wheat and barley. Dixon speculated that farmers often use alfalfa for rotation, and higher water prices may cause them to use wheat and barley, which have lower water requirements (as shown in Table 2.1).

Table 2.7

Relation of Cropping Pattern and Marginal Cost of Water
(Percentage of Observed Farmer-Crop Pairs)[a]

	Marginal Water Cost ($/acre-foot)		
	<= 20 (N=36)	20 to 45 (N=39)	>45 (N=22)
Field crops	72	75	46
Cotton	39	41	32
Alfalfa	33	21	9
Grains	0	13	5
Fruits and nuts	28	24	32
Almonds	22	18	18
Grapes	3	3	0
Citrus	3	3	5
Deciduous trees	0	0	9
Vegetables	0	3	23
Total	100	100	100

SOURCE: Dixon, 1988.

[a]The distribution of farmer-crop pairs is examined rather than simply the percentage of farmers growing particular crops because farmers often operate in multiple irrigation districts that have different water prices.

NOTE: N is the number of farmer-crop pairs in the sample.

The data in Table 2.7 suggest that over an initial range of water cutbacks there will be little change in cropping pattern, but that, once cutbacks become sufficiently large, farmers will switch, over the long run, from field crops to vegetables. Taken together, the studies suggest that water supply cutbacks would cause little change in the preponderance of fruit and nut crops and that reductions in field crops would be concentrated among the high-water-using crops such as alfalfa.

---

[32]Kern County is in the southern San Joaquin Valley.

Irrigation Efficiency. Caswell and Zilberman (1985) found that irrigation technology used for perennial crops (tree and vine crops) varied importantly with water price in the San Joaquin Valley. They found that, as water price rose, sprinkler and drip irrigation became far more common than traditional irrigation (flood and furrow). In an examination of national farm-level data, Negri and Brooks (1990) found that water cost was a statistically significant determinant of irrigation technology choice, but that other farm characteristics, such as the water-holding capacity of the soil and climate, were far more important. The water costs in their dataset were low compared with prices in many parts of the San Joaquin Valley, however; and the sensitivity may be much greater at higher prices.[33]

Dixon (1988) also found a significant and steady shift away from flood and furrow irrigation to sprinkler and drip irrigation as water price rose in Kern County. As shown in Table 2.8, flood and furrow each drop from nearly 50 percent of the farmer-crop observations with marginal water costs less than $20 per acre-foot to less than 20 percent for those observations with water costs greater than $45 per acre-foot. Sprinkler and drip also rise dramatically as price increases.

Table 2.8

Relation of Irrigation Technology and Marginal Cost of Water
(percentage of observed farmer-crop pairs)

	Marginal Water Cost ($/acre-foot)		
	<= 20 (N=36)	20 to 45 (N=39)	>45 (N=22)
Flood and furrow	92	75	36
Sprinkler	8	22	50
Drip	0	2	14
Total	100	100	100

SOURCE: Dixon, 1988.
NOTE: N is the number of farmer-crop pairs in the sample.

The strong relationship between irrigation technology and water price suggests that farmers will move from flood and furrow irrigation to sprinkler and drip irrigation when faced with permanent water cutbacks.

Farmer Response to Surface Water Reductions in the San Joaquin Valley Between 1987 and 1992

Farmer response to the water supply reductions between 1987 and 1992 likely represents the short-run response to reductions that were viewed in part as temporary and in part as permanent. As discussed in Section 1, the bulk of the reductions were due to the drought; however, some of the reductions in 1992 were due to the National Marine Fisheries

---

[33]In Negri and Brook's dataset, mean water cost was $16 per acre-foot with standard deviation $10.

regulations, and farmers may have viewed some part of the NMFS reductions as permanent. What is more, farmers may have anticipated the cutbacks during the drought to continue as new regulations were promulgated. Below, we review several studies on the impact of surface water reductions during the drought.

Procurement of Supplemental Water Supplies. It is widely accepted that groundwater pumping increased a great deal in the San Joaquin Valley during the drought (Archibald et al., 1992; Northwest Economic Associates, 1992; Zilberman, et al., 1994; CDWR, 1991). As will be discussed in Section 3, however, there is much uncertainty about how much groundwater use increased. Zilberman, Howitt, and Sunding (1993) estimate that increased groundwater pumping offset approximately one-third of the decrease in 1991 surface water deliveries statewide.

Farmers in the San Joaquin Valley also bought water from the 1991 Drought Water Bank. There has not been detailed analysis of how this water was used, but most suspect that it was used on perennial crops. The water was expensive ($175 per acre-foot plus transportation costs from the Delta) and bought by fewer farmers than many had expected.

Crop Mix and Crop Fallowing. Northwest Economic Associates (1992) and CDWR (1991) report that a great deal of acreage was fallowed during the drought. Northwest estimated that approximately 250,000 acres were fallowed in the San Joaquin Valley during 1991 (approximately 5 percent of the total), and CDWR estimated that 450,000 acres were fallowed throughout the state. In six case-study irrigation districts spread throughout the Central Valley, Archibald et al. (1992) found a 13 percent drop in acreage harvested between 1990 and 1991. The amount of acreage fallowed suggests that groundwater was either not available, too expensive, or could not be pumped in large enough quantities to fully offset the reductions in surface water deliveries. It also suggests that the drought did indeed have an impact on San Joaquin Valley agriculture.

According to Northwest Economic Associates (1992), cotton, grain, and alfalfa were the crops mainly fallowed, and Archibald et al. found that cotton was the main crop fallowed. Archibald et al. (1992, p. 5-60) observed that fallowing was concentrated in annual, as opposed to perennial, crops.

Archibald et al. also suspect that there may have been some shift in the desired crop mix, suggesting that tomatoes replaced cotton in some areas (1992, pp. 6-60, 6-15). The authors point out, however, that a general trend toward fruits and vegetables and away from grain and cotton preceded the drought (p. 5-14), which makes it difficult to isolate the effect of water cutbacks from other factors. Northwest Economic Associates (1992, p. 23) also suspect that changes in water availability and cost during the drought caused some shift from cotton and grains to vegetables, but conclude that other factors were primarily responsible.

Irrigation Efficiency. Based on a survey of irrigation districts, Zilberman, Howitt, and Sunding (1993) found that there was a significant growth in drip and sprinkler irrigation between 1987 and 1991 throughout the state, and they attribute the growth in part to the drought. Archibald et al. (1992, p. 6-16), on the other hand, observed that few producers in their case-study districts invested in more efficient irrigation technologies in 1990 or 1991. This discrepancy may in part be due to the small sample size in Archibald et al., or the fact that Zilberman, Howitt, and Sunding present data for the entire state and not just the San Joaquin Valley. Some uncertainty does remain, however, on the effect of water supply reductions during the drought on the choice of irrigation technology.

Even though there may be uncertainty on the extent to which new irrigation technologies were adopted, there is general agreement that water supply reductions during the drought induced widespread improvements in irrigation management. Zilberman, Howitt, and Sunding (1993) note the increased use of irrigation scheduling services (such as the California Irrigation Management Information System) and cited one irrigation district manager who said that farmers who continued to use furrow irrigation reduced their per acre water use by more than 10 percent without significantly affecting yields. Archibald et al. (1992, p. 6-16) note efficiency enhancing changes such as night irrigation, laser leveling, and shorter furrows. CDWR (1991, p. 19) also asserts that there were improvements in irrigation efficiency, but does not provide detail on how they occurred.

Studies by Archibald et al. and Northwest examined whether farmers reduced water application rates enough during the drought to reduce yields. Archibald et al. (1992, pp. 5-74) reported lower leaching rates in some of their case study districts but no apparent reductions in yields. They did note, however, that inadequate leaching over the long run may cause a decline in yields as salts concentrate in the soil. Northwest Economics Associates (1992, p. 16) concluded that water supply reductions caused reduced yields on approximately 13 percent of the irrigated acreage in the San Joaquin Valley. However, their findings were based on a survey of irrigation district managers, and it is not clear how the irrigation managers came up with their estimates.

Employment and Other Farm Inputs. Monthly drought bulletins issued by the California Employment Development Department (EDD) (1991) suggest that the drought had only minimal impact on agricultural activity both in the state and in the San Joaquin Valley. Between March and August, 1991, only 402 individuals filed drought-related unemployment claims. Between 3 and 6 percent of 3,600 agricultural employers surveyed reported that the drought had some impact (positive or negative) on their operation, and, what is more, depending on the month only 22 to 55 percent of these employers actually laid off employees or hired fewer employees than normal.[34] Data on drought-related unemployment claims likely understate negative drought impacts because an agricultural worker who is laid off may not file an unemployment claim or, even if he or she does, may not attribute his or her claim to drought.[35] Be that as it may, the EDD data provide little evidence that water supply cutbacks in 1991 had major effects on employment.

---

[34]The vast majority of the remaining employers only *anticipated* that the drought would have an impact on their operations.

[35]This may particularly be the case among California's highly mobile, poorly educated agricultural labor force.

Several analysts have calculated employment effects of water supply reduction in the San Joaquin Valley by multiplying crop-specific average labor requirements per acre by estimates of the number of acres fallowed (see Archibald and Kuhnie, 1994; Northwest Economic Associates, 1992). Based on a 13 percent drop in acreage harvested across their case study districts, Archibald and Kuhnie (1994), for example, predicted a 6 percent drop in employment.[36] There may be many reasons that farmers do not reduce their labor force according to such fixed prescriptions. For example, as mentioned above, farmers may desire to maintain their labor force and thus lay off as few employees as possible in the face of temporary water reductions.

Summary of Empirical Studies on Farmer Response to Surface Water Reductions

The evidence on farmer response to varying water costs reinforces our expectation that farmers will change cropping patterns and irrigation technologies over time in response to *permanent* water cutbacks. As summarized in Table 2.9, there is some evidence that, once cutbacks become sufficiently large, farmers will move away from high-water-using and low-capital- and low-labor-intensive crops (particularly alfalfa). Consistent with expectations, there is also evidence that they will adopt irrigation technologies with higher irrigation efficiencies.

Existing studies of farmer response during the drought suggest that crop fallowing is a primary response to temporary water supply cutbacks and short-run response to permanent water cutbacks. Water reductions during the drought likely had both temporary and permanent components—although it is reasonable to think that the temporary component dominated. The crops fallowed appear to be the same crops that might be phased out over time in response to permanent water cutbacks. Consistent with our conceptual discussion of farmer response, there also is not much evidence that farmers are able to rapidly adopt new irrigation technologies in response to temporary cutbacks. However, significant improvements in irrigation management may have occurred quickly.

The analysis of farmer response to water reductions is based on fragmented data with many gaps. In the next section we attempt to add to the understanding of how farmers respond to water supply reductions by examining changes in county-level data on water use, cropping pattern, yields, agricultural employment, and other measures of agricultural activity during the 1986–1992 drought.

---

[36]The less-than-proportionate drop in labor was because crops that use fewer inputs (field crops) were primarily the ones that were fallowed.

Table 2.9

Summary of Empirical Evidence of Farmer Response to Surface Water Reductions

Area	Long-Run Response to Permanent Reductions	Response During 1986–1992 Drought
Use of alternative water supplies	Not addressed	Increased groundwater pumping; purchases of supplemental surface water
Desired crop mix	Less alfalfa more wheat and barley; fewer field crops and more vegetables	Some weak evidence on shifts from field crops to vegetables
Crop fallowing	Not addressed	Substantial fallowing of field crops
Irrigation technology	Sprinkler and drip replaces flood and furrow	Conflicting opinions on adoption of sprinkler and drip
Irrigation management	Not addressed	Widespread improvements
Crop stressing	Not addressed	Conflicting opinions
Employment and use of inputs	Not addressed	Empirical data suggest minimal impact; models project 6% employment drop

3. THE IMPACT OF WATER SUPPLY CUTBACKS DURING THE DROUGHT ON FARMERS IN THE SAN JOAQUIN VALLEY: NEW EMPIRICAL EVIDENCE

This section presents new empirical evidence on the effect of water supply reductions during the 1987–1992 drought on San Joaquin Valley agriculture. We discuss how water use and various measures of agricultural activity changed between 1985 and 1992 in two counties in the southern San Joaquin Valley where water reductions are thought to have had a significant impact. To help isolate changes caused by water supply reductions from changes caused by other factors, we compare changes in the two southern counties with those in three counties in the northern San Joaquin Valley where water use is thought to have changed little during the same period. The county is the unit of analysis because the county is the most disaggregated level at which much economic and agricultural data are available.

Following a brief description of the general economic backdrop against which water supply reductions took place, we describe the counties selected for the study. We then examine how various measures of agricultural water use changed during this period, evidence on whether these reductions were considered temporary or permanent, and finally the concomitant changes in agricultural activity.

ECONOMIC BACKDROP FOR WATER SUPPLY REDUCTIONS DURING THE DROUGHT

The drought took place during a period of when there were many changes in both the agricultural economy and the economy as a whole. Figure 3.1 presents the recent trends of the California economy. After the 1981–1982 recession, real personal income and non-agricultural employment grew steadily in California through the remainder of the 1980s. Recession caused a downturn in both real personal income and non-agricultural employment in 1991 and weak recovery in 1992. As will be discussed below, this economic downturn coincides with the most severe water supply reductions during the drought.

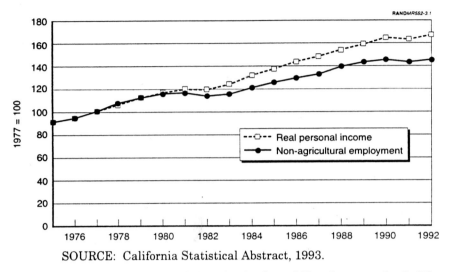

SOURCE: California Statistical Abstract, 1993.

**Figure 3.1—Personal Income and Non-Agricultural Employment in California
(adjusted for inflation and indexed to 1977)**

In the late 1970s and early 1980s, California agriculture was also doing very well. As evidenced by Figure 3.2, net farm income was rising. Land prices were also rising, and farmers took on debt to finance often speculative land purchases (see Figure 3.3). Falling farm prices and escalating debt, however, pulled agriculture down with the rest of the economy in 1982 and 1983. California agriculture did recover during the second half of the 1980s, and farmers were able to reduce their debt burden. Several good years and declining debt meant that the farm economy was in relatively good shape when both water reductions and the economic downturn hit in the beginning of the 1990s.

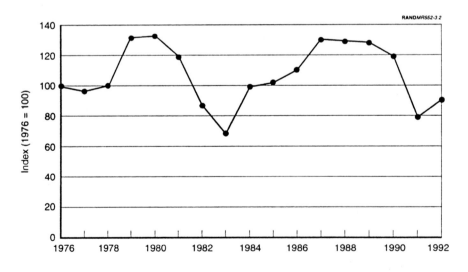

SOURCE: California Statistical Abstract, 1993

**Figure 3.2—Net Farm Income in California
(adjusted for inflation and indexed to 1976)**

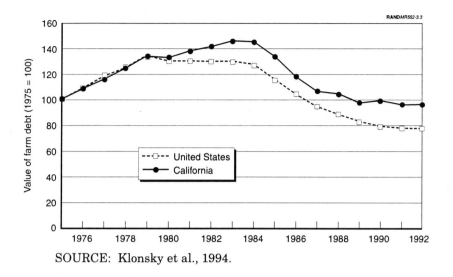

SOURCE: Klonsky et al., 1994.

**Figure 3.3—Farm Debt for the United States and California
(adjusted for inflation and indexed to 1975)**

The drought took place in the midst of a steady change in the relative prices of fruit, vegetable, and field crops. As shown in Figure 3.4, prices of such field crops as feed grain, hay, and cotton have steadily declined relative to those of fruits and vegetables since the mid-1970s. These changes presumably induce shift from field crops to fruits and vegetables—independent of any change in water supplies.

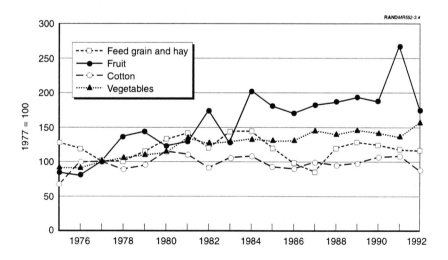

SOURCE: U.S. Department of Agriculture, 1994, p. A-3, and U.S. Department of Commerce, 1986, p. 647.

**Figure 3.4—National Index of Crop Prices
(adjusted for inflation and indexed to 1976)**

During the drought there was also a serious freeze in the southern San Joaquin Valley. The December 1990 freeze severely affected citrus and broccoli yields in Fresno and Kern Counties. The freeze was independent of the drought. In the analysis that follows, we will attempt to control for the effects of the freeze when possible.

COUNTY SELECTION AND CHARACTERISTICS

Data on agricultural water use collected for this study (to be discussed below) suggested that water use fell considerably during the drought in parts of the southern San Joaquin Valley but not a great deal in the northern San Joaquin Valley. Project resources allowed us to collect detailed information on a limited number of counties, so we decided to focus our attention on the two largest counties in the southern San Joaquin Valley—Fresno and Kern (see Figure 3.5). In the analysis below, we refer to these counties as the *impact counties*. We chose three counties in the northern San Joaquin Valley for the purpose of comparison—Merced, San Joaquin, and Stanislaus. We refer to these counties as the *control counties* because they help us to us identify, or control for, changes in agricultural activity that were caused by factors other than the drought. Such changes include changes in the costs of inputs to crop production, crop prices, weather, and crop pests and viruses. For example, as will be discussed above, a gradual rise in fruit and vegetable prices relative to field crops may be causing a general movement away from field crops and into fruits, nuts, and vegetables independent of the drought. We now briefly describe the counties and discuss whether the three northern counties are in fact good controls for the two southern counties.

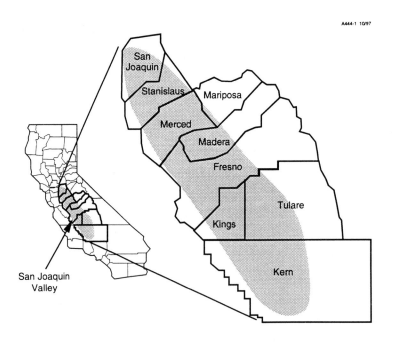

Figure 3.5—Map of the San Joaquin Valley

Characteristics of the Impact and Control Counties

Table 3.1 presents descriptive information on the five counties for 1985—the first year of the period examined. Additional data are included in Appendix A. Crop production (as opposed to livestock and poultry production) accounted for 75 percent or more of total farm revenue in the impact counties. In contrast, crop production accounted for less than half of farm revenues in two of the three control counties. The breakdown of harvested acreage between field crops, fruit and nut crops, and vegetables was similar across the five counties in 1985. As shown in Table 3.1, acreage in field crops ranged from 59 to 73 percent of harvested acreage, and the percentage in fruits and nuts varied from 20 to 30 percent. Vegetables accounted for between 7 and 12 percent of harvested acreage. Even though the breakdown by crop category was similar across counties, the primary crops grown in each category varied to some extent. The primary field crops in both the impact counties were cotton, alfalfa, and wheat. Irrigated pasture (related to the predominance of livestock and poultry) replaces cotton as the top field crop in two of the three control counties. Fruit and nut crops, grapes, almonds, and walnuts were heavily represented in all five counties. Citrus crops, however, were more heavily represented in the impact counties. The predominance of tomatoes is a unifying characteristic in 4 of the 5 counties. The mix of vegetables in Kern County, however, is different from the other four.

The penultimate group of rows in Table 3.1 reports 1985 water use as estimated by the Central Valley Project Environmental Team (to be discussed below). Water use per harvested acre ranges from 2.9 and 3.5 acre-feet per acre in the impact counties. There is more variation in the control counties—water use ranges from 2.2 to 3.9 acre-feet per acre. Surface water accounts for the bulk of water in all the counties.

The final group of rows in Table 3.1 presents average prices of surface water and groundwater in the five counties.[37] In all counties, surface water is cheaper than groundwater, suggesting that groundwater is the marginal water source. Both surface water and groundwater are more expensive on average in the impact counties than in the control counties, with the highest prices in Kern County.

---

[37]The water prices are taken from the Central Valley Production Model (discussed in Section 4). The surface water price is the weighted average of Central Valley Project contract water and State Water Project water, with the weights determined by the use of each type of water. Prices of local surface water are not included in the average.

Table 3.1

Characteristics of Impact and Control Counties in 1985

	Impact Counties		Control Counties		
	Fresno	Kern	Merced	Stanislaus	San Joaquin
Total farm revenue ($billions)[a]	2.065	1.199	0.801	0.748	0.700
Crop production (percent)	75	93	45	38	68
Livestock and poultry (percent)	25	7	55	62	32
Acreage harvested (1000s)[a]	1,397	821	487	408	538
Field crops (percent)	61	64	73	59	64
Fruit and nuts (percent)	23	26	20	30	25
Vegetables (percent)	11	9	7	11	12
Primary field crops[a]	Cotton	Cotton	Pasture	Pasture	Corn
	Wheat	Alfalfa	Cotton	Silage	Alfalfa
	Alfalfa	Wheat	Alfalfa	Beans	Sugar-beets
Primary fruit and nut crops[a]	Grapes	Grapes	Almonds	Almonds	Grapes
	Almonds	Almonds	Grapes	Walnuts	Almonds
	Oranges	Pistachios	Walnuts	Grapes	Walnuts
Primary vegetable crops[a]	Tomatoes	Potatoes	Tomatoes	Tomatoes	Tomatoes
	Melons	Carrots	Melons	Beans	Asparagus
	Garlic	Onions	Potatoes	Peas	Melons
Total water use (millions of acre-feet)[b]	4.018	2.870	1.692	1.582	1.159
Acre-feet per harvested acre	2.88	3.50	3.47	3.88	2.15
Surface water (percent)	63	58	64	68	56
Groundwater (pecent)	37	42	36	32	44
Water Prices ($/acre-foot)[c]					
Surface water[d]	22	35	19	18	14
Groundwater	40	63	22	19	21

[a]County agricultural commissioner reports. Primary crops are top 3 crops by acreage in each crop category.

[b]Data provided by Central Valley Project Environmental Team.

[c]Prices taken from Central Valley Production Model (discussed in Section 4).

[d]Average of CVP contract and SWP prices. Excluded from average are CVP supplemental water and local surface water prices.

How Adequate Are Merced, Stanislaus, and San Joaquin as Controls?

Ideally we would like to compare agricultural activity in the impact counties had there been no surface water supply reductions with agricultural activity during the drought. Of course we cannot observe the former. Merced, Stanislaus, and San Joaquin Counties are good controls to the extent that the changes in agricultural activity between 1985 and 1992 are similar to those that would have occurred in the impact counties had there been no surface water reductions in the impact counties.

Several factors make Merced, Stanislaus, and San Joaquin Counties good controls. They are geographically close to the two impact counties and thus should, on the whole, experience similar weather fluctuations. Their proximity should also guarantee that changes in input and output prices are similar. The similarity of the division of acreage between field, fruit and nut, and vegetable crops suggests that farmers face similar production opportunities in the impact counties as in the control counties. Finally, as we will see shortly, even though surface water supplies in the control counties fell during the drought, overall water use appears to have changed little. Groundwater prices are not a great deal higher than surface water prices in the control counties, which as discussed in Section 2, suggests that there may be little change in farm operations.

There are also several significant differences between the impact and control counties—some of which are important in evaluating the suitability of the controls and others which are not. Livestock and poultry account for a larger percentage of farm revenue in the control counties, but this is not particularly important because we analyze the livestock and poultry sector separately from the crop production sector and will look for shifts between the two sectors.

One potentially important difference between the impact and control counties is in the specific crops grown in each crop category. For example, cotton is the dominant field crop in the impact counties, and some of the changes in the impact counties between 1985 and 1992 may be attributable to peculiarities of the cotton market or advances in cultivation practices for cotton relative to other crops. Also, as noted above, the December 1990 freeze was felt largely in the impact counties. We will attempt to control for this difference in the analysis below.

Although we cannot be certain that the observed changes in agricultural activity in Merced, Stanislaus, and San Joaquin Counties are very close to the changes that would have occurred in Fresno and Kern Counties had there been no surface water cutbacks, we think the comparison between the impact and control counties worth making. We will not, however, infer the impact of water supply reductions by mechanically comparing the two sets of counties; rather, we will inform the comparison with a general understanding of the context in which the water supply reductions took place. In comparing the two sets of counties, we will also keep in mind that we have small sample sizes. Thus we will look for consistent patterns across time and across different measures of agricultural activity to help us feel comfortable that we are observing more than random fluctuations in the data.

HOW MUCH DID WATER USE CHANGE DURING THE DROUGHT?

There is a great deal of uncertainty about how agricultural water use actually changed during the drought. Surface-water diversions to Central Valley Project and State Water Project contractors are generally well monitored and the data readily available. Diversions by users with appropriative water rights on local rivers are also generally monitored, but the data are usually available only from local water districts. Riparian and groundwater use, in contrast, are very poorly monitored. The lack of data on groundwater use on a regional basis is particularly unfortunate in examining the impact of the drought because farmers in many parts of the San Joaquin Valley are thought to increase groundwater pumping when surface water supplies decline.[38]

Measures of Water Use

We use estimates of annual surface water deliveries in the San Joaquin Valley between 1985 and 1992 developed by a Bureau of Reclamation research effort (the Central Valley Project Environmental Team—CVPET).[39] Due to uncertainty regarding groundwater pumping, we use two different estimates of groundwater pumping: (1) estimates developed by the CVPET, and (2) estimates derived from crop-specific water application rates.

The estimates of groundwater assembled by the CVPET came from two sources. Data between 1985 and 1990 are taken from a Bureau of Reclamation-sponsored groundwater modeling effort that is presently inactive.[40] The numbers are based in part on a United States Geological Survey study that inferred groundwater pumping from electricity-use data.[41] The estimates for 1985–1990 used here were developed just prior to the suspension of the study in 1992 and were not finalized.[42] The CVPET estimates for 1991 and 1992 were developed independent of the groundwater modeling effort. According to CVPET staff, estimates for Kern and Fresno Counties were based on pumping data collected by local water agencies and are thought to be fairly reliable. The estimates for the control counties appear to be primarily based on cropping pattern, using a method that we now describe.

The second measure of groundwater use is based on crop-specific average water application rates. First, total water use is determined by multiplying the acreage planted in each crop by its average water application rate. Groundwater use is then calculated by subtracting surface water use from calculated total water use. We refer to the resulting

---

[38]Riparian use, in contrast, is likely to be much more stable over different types of water years.

[39]The data are drawn from historic measurements of surface water deliveries maintained by the Bureau of Reclamation, the United States Geological Survey, the California Department of Water Resources, and local districts. The data were provided directly to the authors in 1994.

[40]The project was part of the environmental impact study for the Friant Unit contract renewals.

[41]Most groundwater pumps are electrically driven. Given estimates of depth-to-water and pump efficiency, groundwater pumping can be inferred from electricity use.

[42]Staff who worked on the Bureau-sponsored study emphasize the preliminary nature of the numbers and that they may be revised if the study is reactivated.

total water and groundwater use as *derived* total water use and *derived* groundwater use. We calculated derived total and groundwater use by using the average water application rates adopted by the CVPET in their modeling efforts, which are in turn based on studies done by the California Department of Water Resources.

Most studies of agricultural water use in the San Joaquin Valley rely on water use estimates derived from cropping patterns. The lack of better data may make this necessary, but the derived estimates may not be very accurate. Estimates based on applied water use coefficients that are constant across time are likely to understate changes in water use during a drought. First, as discussed in Section 2, application rates vary by irrigation technology, and implied water use will miss reductions in water use due to switches to drip and sprinkler irrigation during drought. Second, during periods of water scarcity, farmers may pay more attention to irrigation practices and increase irrigation efficiency. Finally, farmers may deficit irrigate their crops during periods of water shortages. Derived water use data in effect assume away many possible farmer responses to water supply reductions. Consequently, estimates based on indicators of groundwater pumping independent of cropping pattern are preferable. The CVPET attempted to produce such estimates, with the apparent exception of groundwater use in the control counties in 1991 and 1992.

Estimates of Water Use

The CVPET data indicate that surface water diversions in Kern and Fresno Counties fell substantially between 1985 and 1992. As shown in the left hand side of Figure 3.6, surface water diversions fell between 1986 and 1987, stabilized somewhat in 1988 and 1989, but then fell rapidly in 1990 and 1991. Surface water deliveries rebounded somewhat in 1992 in the impact counties, but remained depressed. Numerical data on changes in water use in each county are presented in Appendix A.

Both the CVPET and derived groundwater estimates show groundwater use rising in impact counties over this period; however, the two estimates tell different stories about total water use. As will be detailed below, the CVPET estimates show that total water use fell approximately 15 percent in both Fresno and Kern Counties from average use in 1987, 1988, and 1989 (1987–1989) to 1991 and approximately 7 percent from average use in 1987–1989 to 1992. The derived groundwater estimates, in contrast, imply that total water use fell roughly 7 percent between 1987–1989 and 1991 and 6 percent between 1987–1989 and 1992 in both counties.[43]

As shown in Figure 3.7, surface water cutbacks during the drought were much less severe in the control counties, and even appear to have risen slightly in San Joaquin County. The two estimates of groundwater use differ somewhat, but both imply that total water use fell little if at all over the period.

---

[43]The larger reduction in the CVPET estimates than the derived estimates is consistent with the hypothesis that the derived estimates do not accurately capture the responses of farmers during drought.

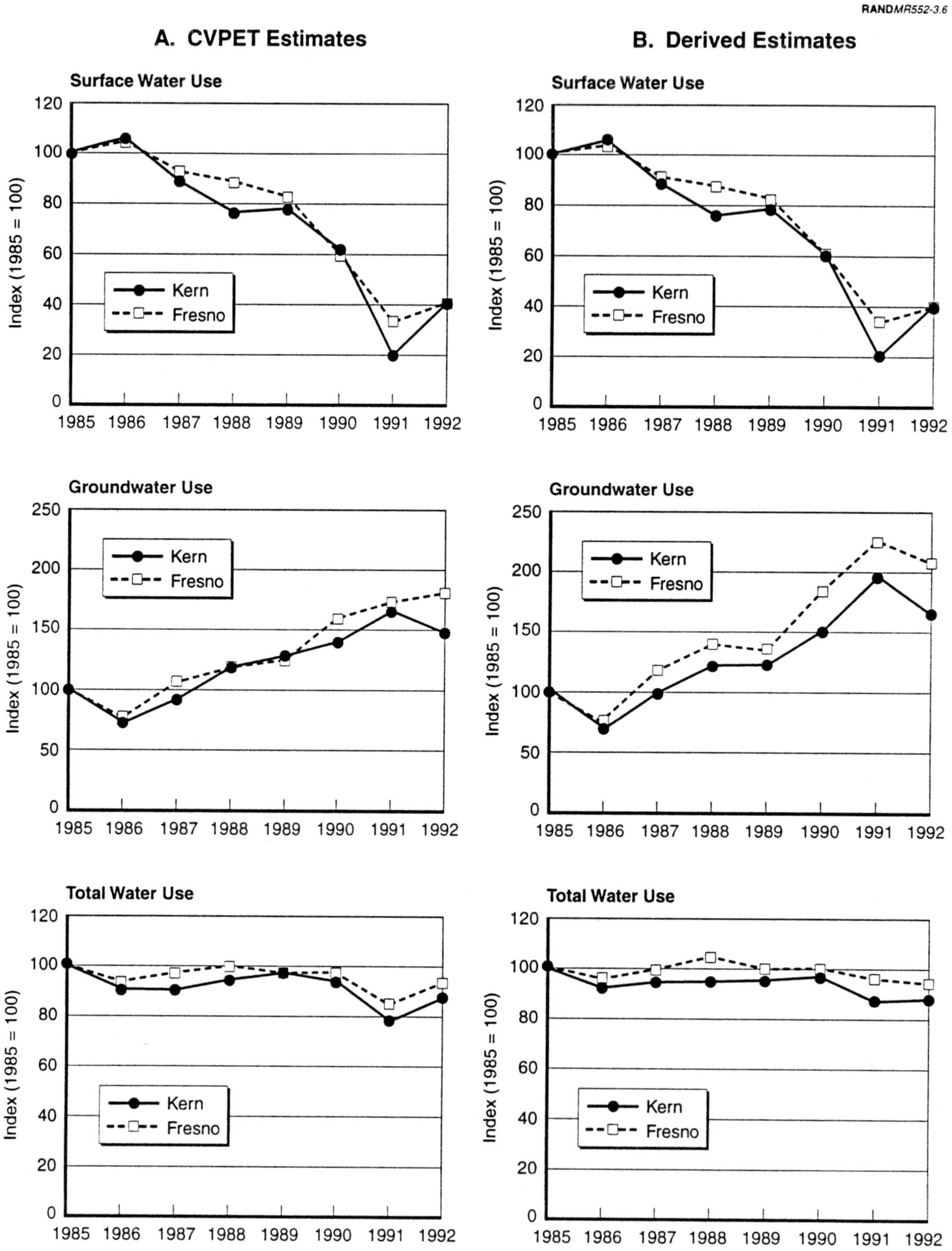

A. CVPET Estimates

B. Derived Estimates

Figure 3.6—Estimates of Water Use in Impact Counties

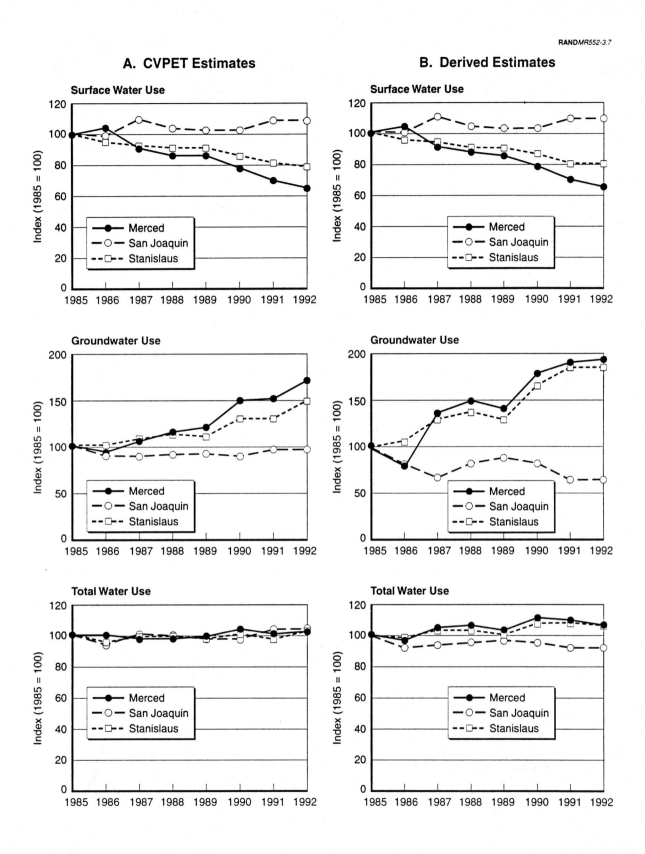

Figure 3.7—Estimates of Water Use in Control Counties

Evaluation

In our analysis, we focus on the change in agricultural activity and water use over three periods: (1) between 1985–1986 and 1987–1989; (2) between 1987–89 and 1991; and (3) between 1987–1989 and 1992. These periods are based on the patterns of surface water use observed in the impact counties. Separately examining the change between 1987–1989 and 1991 and then between 1987–1989 and 1992 allows us to better isolate potentially confounding factors such as the December 1990 freeze that may have caused agricultural losses in 1991 independent of the water supply reductions during the drought.

Table 3.2 summarizes that average percentage change in water use in the impact and control counties. The averages are calculated by averaging the percentage changes for each county in the appropriate set of counties. There were large reductions in surface water deliveries in the impact counties during the drought. These reductions were to a substantial extent offset by pumping of higher-priced groundwater.[44] Exactly how much the reductions were offset is open to question, although, given the wealth of groundwater pumping data in Fresno and Kern Counties, we put more stock in the CVPET estimates. It thus seems likely that, in the aggregate, farmers were able and willing to offset the surface water reductions during the first part of the drought but did not in the latter part of the drought. The substantial fall in water use between 1987–1989 and 1991 and between 1987–1989 and 1992 implies that farmers either cut back use because they switched to a higher cost marginal water supply, that they had limited or no access to groundwater, or both.

These percentage declines translate into substantial reductions in the number of acre-feet used. Between 1987–1989 and 1991, surface water use in Fresno and Kern Counties fell 2.43 million acre-feet, and total water use fell 970,000 acre-feet.

The data suggest that the moderate cutbacks in surface water supplies were offset by increased groundwater pumping in the control counties; however, both the CVPET and

Table 3.2

Changes in Water Use in the Impact and Control Counties Between 1985 and 1992
(average percentage change)

	1985–1986 to 1987–1989		1987–1989 to 1991		1987–1989 to 1992	
	Impact	Control	Impact	Control	Impact	Control
CVEPT estimates						
Surface water	–18	–5	–70	–10	–53	–12
Groundwater	32	8	47	20	45	30
Total	0	–1	–15	2	–7	5
Derived estimates						
Surface water	–18	–5	–70	–10	–53	–12
Groundwater	42	24	72	20	52	18
Total	1	3	–7	2	–6	0

---

[44]Our discussion here is of water use in the county as a whole. Some farmers in the county may have purchased supplemental surface water from either within or outside the county.

derived estimates are based on crop water use coefficients for 1991 and 1992, whose limitations we have discussed above. Nevertheless, given the widespread availability of groundwater in the control counties at a cost that is similar to surface water, we think it reasonable to conclude that farmers in the control counties did completely offset the surface water reductions with increased groundwater pumping.

WERE WATER SUPPLY REDUCTIONS PERCEIVED AS TEMPORARY OR PERMANENT?

As discussed in Section 2, farmer response to water supply reductions will depend in part on whether farmers view them as temporary or permanent. They may view part of the reductions as permanent if some part of the reduction was due to an actual change in policy (such as the NMFS regulations) or reflective of policy changes to come.

Because agricultural land values are based on the capitalized value of the earnings potential of the land over a long period of time, declining land values are one indicator that farmers and other investors expect long-term reductions in water supplies.[45] Temporary cutbacks due to expected weather fluctuations, in contrast, should have little effect on land values.

In this subsection, we compare how land values changed in the impact and control counties during the drought.[46] We expect to see little change in land values in the control counties *due to changes in water availability* during this period because farmers in the control counties generally hold very senior water rights. Water rights in the impact counties are usually junior to those in the control counties, and it is likely that regulatory water reductions will be focused on senior water rights holders. Background on the determinants of land prices and a more detailed discussion of land prices in the San Joaquin Valley are presented in Appendix B.

Land Value Changes in the Impact and Control Counties

As shown in Table 3.3, inflation-adjusted land prices in both sets of counties fell between 1982 and 1987 and then rose between 1987 and 1992. This pattern mirrors changes in agricultural land values in California and the United States as a whole during this period and illustrates that land prices are determined at least in part by larger economic trends.

Even though land prices in both sets of counties fell between 1982 and 1987, they fell much faster in the impact counties than in the control counties on average. Discussions with farm experts suggest, however, that there was little or no awareness in the farming community prior to 1987 that environmental regulations or the reallocation of water from agricultural to urban uses would decrease agricultural water supplies. Thus, the greater

---

[45]Increasing variability in water supplies might also decrease land values if investors are risk averse, but we ignore this possibility here.

[46]Note that land values decline as the perceived likelihood of water cutbacks increases—which may be well in advance of any actual water cutbacks.

decline in land prices in the impact counties relative to the control counties between 1982 and 1987 was likely due to factors other than water supply reductions. One prime candidate is the increased difficulty farmers in Fresno and Kern Counties had disposing agricultural drainage water once selenium contamination in Kesterson Reservoir was linked to agricultural drainage in the mid-1980s.

Table 3.3

Land Values in the Impact and Control Counties
Between 1982 and 1992
(1985 dollars)

	1982	1987	1992
Impact counties			
$/acre	2,532	1,476	1,657
% change	—	−41	10
Control counties			
$/acre	2,994	2,349	2,657
% change	—	−22	13
Difference			
$/acre	−462	−873	−1,000

SOURCE: 1992 Census of Agriculture.

In percentage terms, land values rose only slightly more slowly in the impact counties between 1987 and 1992, but the absolute difference in land values continued to grow, reaching $1,000 per acre by 1992 in 1985 dollars. Part of the widening difference in land values between the impact and control counties may reflect the realization that surface water supplies would likely be permanently reduced in the impact counties.[47]

The decline in land value in the impact counties relative to the control counties is consistent with anecdotal information provided by farmers and county assessors interviewed during the course of this study in the San Joaquin Valley. There is widespread belief that expected regulatory cutbacks as well as reallocation of water from the agricultural to the urban sector had caused land values in the impact counties to fall. In an anecdote typical of many others, one observer described two closely situated pieces of land that had different water priorities. He estimated that land with the more junior water right had declined 50 percent in value in recent years while the value of land with the more senior water rights had remained unchanged.

The divergence of land values in the impact and control counties between 1987 and 1992 suggests that farmers expected permanent water cutbacks. Thus, some of their

---

[47]Analysis on the relation between the timing of regulatory announcements and changes in land values would allow a better understanding of how much a change in water delivery expectations was responsible for the change in land values. Such an analysis would require more frequent data on land values, however.

adjustments during the drought may have been short-term adjustments to permanent, as opposed to merely temporary, water supply reductions.[48]

HOW DID VARIOUS MEASURES OF AGRICULTURAL ACTIVITY CHANGE?

In this subsection, we examine whether we can detect any reduction in agricultural activity due to water supply reductions during the drought.

Acreage Harvested

As shown in Table 3.4, overall acreage harvested in the impact and control counties did not change much in the first years of the drought. In later years, however, water supply reductions in the impact counties appear to have caused a decline in total acreage harvested. Total acreage harvested fell 10 percent between 1987–1989 and 1991 and 9 percent between 1987–1989 and 1992 in the impact counties while actually rising 4 percent in the control counties.[49]

The changes in acreage harvested varied importantly by crop category. Table 3.4 shows that there was a rapid increase in vegetable acreage during the drought: Vegetable acreage rose 21 percent on average between 1985–1986 and 1987–1989 in the impact counties, another 21 percent between 1987–1989 and 1991, and demonstrating that this was not a fluke, 16 percent between 1987–1989 and 1992. Acreage of vegetables also rose in the control counties but not nearly as fast.

The rapid increase in vegetable acreage in the impact counties relative to the control counties suggests that the water supply cutbacks during the drought induced a shift to vegetables. However, the increases between periods are not closely associated with the percentage changes in total water use or the percentage declines in surface water use, raising questions about whether the growth in acreage of vegetables was due to changes in surface water availability or to other factors.[50]

The declines in fruit and nut acreage in the impact counties were larger than in the control counties during the drought, but the differences are not substantial. What is more, once citrus crops and broccoli are excluded to remove possible effects of the freeze, the differences are even lower (see second set of rows in Table 3.4).

---

[48]The fall in land prices due to permanent water cutbacks represents a real loss to landowners. It is important to note, however, that it was subsidized irrigation water that elevated land prices in the first place (see U. S. House of Representatives, 1994, p. 61).

[49]To calculate the percentage changes for acreage harvested in Table 3.3, the acreage harvested in each county for the years in each of the periods is calculated. For example, the 1985–1986 acreage harvested for Fresno County is the average of acreage harvested in 1985 and 1986. Then the percentage changes for each county between periods are calculated and averaged separately for the impact and control counties.

[50]It is also possible that the shift to vegetables represents a gradual adjustment to the changes in water availability.

Table 3.4

Changes in Acreage Harvested and Yield in Impact and Control Counties Between 1985 and 1992

| | 1985–1986 Average | | Average Percentage Change | | | | | |
| | | | 1985–1986 to 1987–1989 | | 1987–1989 to 1991 | | 1987–1989 to 1992 | |
	Impact	Control	Impact	Control	Impact	Control	Impact	Control
Acreage harvested (thousands)[a]	1,013	470	1	1	−10	4	−9	4
Field crops	629	306	−1	−1	−18	4	−15	7
Fruits and nuts	270	118	−2	2	−4	1	−3	0
Vegetables	114	46	21	6	21	12	16	3
Acreage harvested excluding citrus and broccoli (thousands)[a]	1,013	470	1	1	−10	4	−9	4
Field crops	629	306	−1	−1	−18	4	−15	7
Fruits and nuts	270	118	−3	2	−2	1	−2	0
Vegetables	114	46	21	6	22	12	16	3
Field crop acreage harvested[a] (thousands)								
Lower-value field crops[b]	206	201	−22	−4	−33	2	−31	8
Higher-value field crops[c]	423	105	10	5	−13	9	−10	6
Crop yield excluding citrus and broccoli (1985–1986=100)	99	93	3	10	1	−2	10	4
Field crops	97	97	0	3	2	2	10	9
Fruits and nuts	99	81	14	31	−4	−12	11	−4
Vegetables	104	101	−4	0	3	0	3	5

[a] County agricultural commissioner data.

[b] Crops with gross revenue per acre less than $500. Examples include corn, wheat, and irrigated pasture.

[c] Alfalfa, dry beans, cotton, and sugar beets.

Field crop acreage changed little in the first years of the drought, but then dropped nearly 18 percent in the impact counties between 1987–1989 and 1991 and 15 percent between 1987–1989 and 1992. In contrast, there was actually a rise in field crop acreage in the control counties between 1987–1989 and 1991 and between 1987–1989 and 1992. Even though some of the acreage in field crops may have been converted to vegetables, the large baseline difference between field crops and vegetable acreage means that most of the decline was due to fallowing. Even if the entire difference in the growth of vegetable crops between the impact and control counties between 1987–1989 and 1991 (9 percent) was converted from field crops, the decline in field crop acreage changes only slightly—from 18 to 16 percent between 1987–1989 and 1991.

Table 3.4 shows that throughout the drought there was growth in the acreage of high-value field crops (alfalfa, dry beans, cotton, and sugar beets) relative to lower-value crops (such as corn, wheat, and irrigated pasture) in the control counties.[51] The differential was much greater in the impact counties, suggesting that water supply cutbacks caused a shift from lower- to high-value field crops. Again, however, the steady shift throughout the period raises suspicions that factors other than water supply cutbacks are responsible. The data also indicate that farmers fallowed both higher- and lower-value crops in response to water cutbacks and that lower-value crops remained in the rotations of some farmers.

Several of these findings are consistent with previous empirical research reviewed in Section 2 on the impact of the 1987–1992 drought on crop fallowing and crop mix: little change in fruit and nut acreage and the fallowing of field crops. The findings strengthen the evidence that water cutbacks induce shifts to vegetables and from low-value to high-value field crops, although uncertainty about whether other factors are driving these changes remain. The analysis provides evidence that farmers fallow both low- and high-value field crops in response to water cutbacks.

Our analysis adds information on the response to different sized water reductions. It suggests that with the water infrastructure in place during the drought, farmers are able to accommodate moderate declines in surface water deliveries without much change in acreage harvested—at least in the short run. The 15 to 20 percent reductions in surface water in the control counties between 1985–1986 and 1991, or 1992, had little effect, as did the 20 percent decline in surface water in the impact counties between 1985–1986 and 1987–1989. The large additional reductions in the impact counties between 1987–1989 and 1991, or 1992, did have significant effect on acreage harvested, however.

Irrigation Efficiency

Quantitative information on irrigation technologies and management practices by county are not readily available, and we did not assemble data on changes in irrigation technology and management practices during the drought. However, comparison between

---

[51]Low-value field crops are defined as those crops whose gross revenue per acre is less than $500.

the CVPET and derived estimates of total water use provides some indirect evidence on changes in irrigation technology and management practices.

Recall that the derived estimate measures water use is calculated by applying assumed crop-specific water application rates to the cropping pattern. The CVPET estimate, in contrast, attempts to measure actual water use. Thus, the difference between changes in the two is a measure of the change in water use per acre, controlling for cropping pattern. This difference in water use per acre may reflect changes in irrigation technology, irrigation management practices, or deficit irrigation. To search for any changes in irrigation management, we first look for evidence that farmers deficit irrigated crops enough to reduce yield. We then use these findings to help interpret differences between CVPET and derived estimates of water use.

Changes in Crop Yield During the Drought. We compared how yields changed during the drought in the impact and control counties relative to 1985. For each county, we indexed annual crop yields reported by county agricultural commissioners to their 1985 values (1985=100). We then calculated the changes in yield for each county, weighting each crop by its irrigated acreage in 1987–1989.

The last set of rows in Table 3.4 show that there was little difference between the change in yields in the impact and control counties for both field crops and vegetables. Indeed the yield on field crops jumped significantly between 1987–1989 and 1992 (10 percent), even surpassing the 9 percent jump in the control counties. The 31 percent jump in fruit and nut yields between 1985–1986 and 1987–1989 stands out, but represents a recovery from very low yields in 1986 (the index for 1985–1986 was 81 is the control counties). Overall, the yield on fruit and nut crops did not suffer in the impact counties relative to the controls. The yield index for fruits and nuts was approximately 110 in the impact counties in 1991 and 94 in the control counties. In 1992, the numbers were approximately 125 in the impact counties and 101 in the controls.

The countywide data suggest that water supply cutbacks had little if any effect on yield. Caution is warranted, however, first because the yield numbers appear to be fairly volatile and may be hiding drought effects, and second because the quality of the land harvested may have changed during the drought. If it is the marginal lands that are taken out of production, one might expect overall yield in the impact counties to rise relative to the control counties. Thus the lack of any change in overall numbers could possibly be consistent with crop stressing on the acreage remaining in production.

We think it unlikely that either of these caveats are strong enough to change the interpretation of the county-level yield data.[52] Our conclusion that water supply cutbacks during the drought had little effect on yield is consistent with that of Archibald et al. (1992, p. 5-75), who found little evidence of crop stressing during the drought, and inconsistent

---

[52]Note that lower water application rates during the drought may not have caused reduced yields during the drought, but perhaps at the cost of inadequate salt flushing. Salt buildup over time causes yields to drop.

with Northwest Economic Associates (1992, p. 16), who claimed that there were reduced yields on approximately 13 percent of irrigated acreage in the San Joaquin Valley.

Comparison of CVPET and Derived Water Use. Table 3.5 reproduces the CVPET and derived total water use estimates in the impact counties during the drought.[53] The only evidence of a reduction in water application rates is between 1987–1989 and 1991, and the reduction disappears by 1992.

Reductions in water supply were greatest in 1991, and it may well be that farmers improved irrigation management during that year. The small difference between the CVPET and derived estimates by 1992 suggests that there was little change in irrigation technology (changes are not quickly reversible), so the gap in 1991 may have been due to improvements in irrigation management. Such a conclusion would be consistent with the widespread view that farmers did improve irrigation management during the drought (see Section 2).[54,55] However, the gap between the CVPET and derived estimates in 1991 could also be due to deficit irrigation (although not large enough to affect crop yields), leaving uncertainty about the extent to which farmers did change irrigation practices.

Table 3.5

Average Changes in Water Application Rates in Impact Counties
(percentage)

	1985–1986 to 1987–1989	1987–1989 to 1991	1987–1989 to 1992
CVPET estimates	0	–15	–7
Derived estimates	1	–7	–6
Difference (CVPET – derived)	–1	–8	–1

Value of Agricultural Production

The value of agricultural production is the revenue farmers receive for their products. It is an indicator of the overall size of the agricultural economy because it covers labor payments, purchases of other inputs, and farmer profit. We first examine changes in the value of crop production. We then examine changes in the value of livestock and poultry production.

Value of Crop Production. The value of crop production is the product of acreage harvested, crop yield, and crop prices. For each crop category we analyze the change in value, making reference to each of these components.

---

[53]We do not report data for the control counties because, unfortunately, the 1991 and 1992 CVPET estimates for the control counties are calculated using the same method as the derived estimates.

[54]Our findings suggest that irrigation management improved only during the most severe year of the drought.

[55]It is presumably costly to improve irrigation management, so improvements may make sense only when there are very large water supply cutbacks.

- **Field crops.** Yields changed similarly in the impact and control counties during the drought, and as shown in the first set of rows in Table 3.6, so did field crop prices. The substantial declines in field crop acreage between 1987–1989 and 1991 and 1987–1989 and 1992 show up as a sizable decline in value. Field crop value fell 25 percent in the impact counties between 1987–1989 and 1991 versus 6 percent in the control counties. The gap was smaller but still substantial in 1992.

- **Fruits and nuts.** There was not much change in acreage. Prices and yield moved around quite a bit, but there are no consistent patterns. Thus, once citrus and broccoli are excluded, there is no clear effect of water supply cutback on the value of fruit and nut production.

- **Vegetables.** There was a significant increase in vegetable acreage in the impact counties relative to the controls and not much change in yields. There was a large decline in vegetable prices relative to the controls between 1987–1989 and 1992, but because a similar relative drop did not occur between 1987–1989 and 1991, it seems likely that the 1992 drop was due to factors other than the drought. When crop prices are held constant, the value of vegetable production rose substantially faster in the impact counties than in the controls (see Table 3.6).

During the drought, the value of field crop production thus fell, that of vegetables rose, and that of fruit and nuts moved randomly. Holding crop prices constant (which do not appear to have been affected by the drought) field crop value fell between 18 and 23 percentage points more in the impact counties than in the control counties between 1987–1989 and 1991 or between 1987–1989 and 1992. Given that field crops accounted for approximately one-third of total value in the impact counties at the baseline (1985–1986), this translates into approximately a 6 or 7 percent decline in total value. The 15 to 20 percent increase in vegetable value in the impact counties relative to the control counties offsets this decline by 3 or 4 percentage points.

Table 3.6

Changes in Value of Agricultural Production and Crop Prices in Impact and Control Counties Between 1985 and 1992

| | 1985–1986 Average | | Average Percentage Change | | | | | |
| | | | 1985–1986 to 1987–1989 | | 1987–1989 to 1991 | | 1987–1989 to 1992 | |
	Impact	Control	Impact	Control	Impact	Control	Impact	Control
Crop prices (1985=100)								
Field crops	105	98	1	7	−10	−13	−15	−20
Fruits and nuts	121	137	−6	−5	4	−7	−12	0
Vegetables	101	105	−2	−4	−10	−10	−30	−12
Value of crop production ($million)	1,371	364	7	20	−12	−9	−11	−5
Field crops	417	119	6	7	−25	−6	−17	−9
Fruits and nuts	636	157	4	37	−14	−14	−4	−2
Vegetables	293	84	13	10	12	2	−16	−7
Value of crop production excluding citrus and broccoli ($billion)	1,267	0.364	7	20	−7	−9	−9	−5
Field crops	417	119	6	7	−25	−6	−17	−9
Fruits and nuts	540	157	4	37	−5	−14	1	−2
Vegetables	286	84	13	10	15	2	−15	−7
Value of crop production with constant crop prices ($million)[a]	1,293	348	9	20	−6	0	7	4
Field crops	403	110	5	3	−16	7	−2	16
Fruits and nuts	604	159	8	32	−12	−9	7	−2
Vegetables	264	76	16	16	25	11	25	5
Value of livestock and poultry production ($ million)	275	374	1	−4	1	−2	5	−5

[a]Crop prices held constant at their average values between 1985 and 1992.

The effect of the drought on crop value hinges on whether one believes that the more rapid shift to vegetables in the impact counties was due to the water supply cutbacks or to other factors. If one believes it was due to other factors, then the water supply cutbacks between 1987–1989 and 1991 or 1992 caused a 6 or 7 percent decline in value. If shift to vegetables was due to water supply cutbacks, the decline is half that.

Apparently, random changes in the components of crop value make it difficult to find any effect of the drought on the aggregate value of crop production. Even when crop prices are held constant, the change in total value between 1987–1989 and 1991 is lower in the impact counties than in the controls while it is greater between 1987–1989 and 1992. The large increase in the yield of fruit and nut acreage in the impact counties relative to the controls in 1992 is largely responsible for this reversal—an increase that most likely was not due to the drought.

Value of Livestock and Poultry Production. Table 3.6 shows that the value of livestock and poultry production grew faster in the impact and in the control counties during the drought. Thus, declines in crop value could conceivably be explained in part by shifts between the crop production and livestock and poultry sectors.

Agricultural Employment

California's Employment Development Department (EDD) provided us with data on farm wage and salary employment and payroll by county during the drought. These data are based on the information reported by employers when they pay unemployment taxes for their workforce.

The data on whether water supply cutback affected on-farm crop production employment are ambiguous. The ambiguity is due to the relatively rapid growth in on-farm employment in the control counties between 1985–1986 and 1987–1989 (see Table 3.7). If one views this 6 percent increase as a rebound from unusually low levels in 1985–1986, perhaps related to the low yields in the control counties in 1985–1986, then it is appropriate to compare 1991 and 1992 employment levels with 1987–1989. Employment fell similarly in the impact and control counties between 1987–1989 and 1991 or 1992, suggesting that the water supply cutbacks affected on-farm employment little. If one views the 6 percent increase as a random increase from normal levels, then it is appropriate to compare 1991 and 1992 with 1985–1986. Crop production employment fell 5 percent more in the impact counties between 1985–1986 and 1991 and between 1985–1986 and 1992, suggesting that water supply cutback caused a 5 percent drop in on-farm employment.

Table 3.7

Changes in Agricultural Employment in Impact and Control Counties Between 1985 and 1992

| | 1985–1986 Average | | Average Percentage Change | | | | | |
| | | | 1985–1986 to 1987–1989 | | 1987–1989 to 1991 | | 1987–1989 to 1992 | |
	Impact	Control	Impact	Control	Impact	Control	Impact	Control
Crop-production employment								
Average employment	38,129	10,310	7	12	-2	-3	-3	-9
Annual payroll ($millions)[a]	332	88	6	16	0	2	3	-1
On-farm employment								
Average employment	21,453	6,743	1	6	-6	-6	-7	-7
Annual payroll ($millions)[a]	187	61	4	14	4	2	2	2
Farm services employment								
Average employment	16,675	3,567	14	24	4	2	3	-11
Annual payroll ($millions)[a]	144	27	8	21	-1	4	7	-5
Livestock and poultry employment								
Average employment	2,135	2,343	-6	4	6	2	-3	-2
Annual payroll (millions)[a]	29	33	-5	8	7	4	-3	2
Crop-related food and kindred products employment								
Average employment	3,407	6,412	4	-1	4	-7	5	-2
Annual payroll ($millions)[a]	68	140	7	-5	5	-2	5	1
Total county employment	226,000	115,000	6	10	3	2	8	5

SOURCE: California Employment Development Department unemployment tax database, except total county employment, which is from California Employment Development Department, 1993.

[a]Converted to constant dollars using the gross domestic product deflator (1985=100).

In the absence of data further back in time, we have no basis for choosing between the two interpretations. Employment levels may have been unusually low in the control counties in the base year because of yield problems. On the other hand, employment levels in the control counties in 1991 and 1992 were about the same as in 1985–1986, suggesting we should ignore the increase between 1985–1986 and 1987–1989.[56]

There is no consistent evidence that farm services employment related to crop production (for example, pesticide applicators or crop-harvesting firms) fell more in the impact counties: Even if one views the 24 percent increase in the control counties between 1985–1986 and 1987–1989 as an increase from normal levels of employment, farm service employment rose slower in the impact counties between 1985–1986 and 1991 but faster between 1985–1986 and 1992.[57]

Employment in the food processing and packaging industry does not appear to have been adversely affected during the drought.[58] There is no clear evidence that livestock and poultry employment grew more slowly in the impact counties than in the control counties, nor is there much evidence that total county employment grew more slowly in the impact counties.

To summarize, our county-level data provide no strong evidence that water supply cutbacks during the drought affected employment. It is possible that there was a drop in on-farm crop production employment in 1991 and 1992, and perhaps a decline in farm-services employment in 1991, but more data are required to determine whether this was the case. We thus can do little to resolve the current lack of consensus on the impact of the drought on employment (see Section 2).

It is possible that a rapid increase in vegetables in the impact counties that is unrelated to the drought is masking some of the drought-induced reductions of employment. Unfortunately, data on on-farm crop production employment by crop-type are not available, so we cannot separate the change (likely negative) in field crop or fruit and nut employment from the change (likely positive) in vegetable employment.[59]

---

[56]One might expect seasonal workers to be affected more adversely than full-time workers, but reliable data distinguishing between seasonal and regular workers are not available. California's Employment Development Department produces Report 881-M, which gives annual estimates of the number of seasonal and regular employees in agriculture by county. However, these estimates are not based on survey data and are apparently derived by applying fixed labor requirements for different types of crops to cropping pattern.

[57]It is possible that farm services firms were hurt by water supply reductions in 1991, but were successful in marketing their services in areas less affected by the drought in 1992.

[58]The lack of impact in the food processing and packaging industry is not surprising because only a small percent (5 to 10 percent) of food processing and packaging inputs in the San Joaquin Valley come from the San Joaquin Valley (U.S. EPA, 1994, p. 5-12).

[59]Even if the shift to vegetables was a response to water supply cutbacks, it is important to note that the gainers from increased vegetable production are not necessarily the same as the losers from lower field crop production.

Farmer Income

As argued in Section 2, we expect farmer income to fall during the drought. The most reliable data with which to test this expectation are from the Census of Agriculture. The Census of Agriculture is done only every five years, but, fortunately, it was done in 1992, a year in which there were still large surface water cutbacks in the impact counties.

Table 3.8 shows that net farm cash returns fell 20 percent in the impact counties versus 14 percent in the control counties between 1987 and 1992.[60] This suggests that farmers were indeed hurt by the drought. When interpreting the numbers, the reader should keep in mind first that net cash returns include livestock and poultry as well as crop production, and the change in cash returns may differ across the two sectors. For example, if water supply cutbacks had little effect on the livestock and poultry sector, then the effect on the crop production sector would be larger than for the two combined.[61] Second, it is important to note that changes in crop prices and any exogenous shift to vegetables are included in the numbers and may mask, or exaggerate, the effects of water supply reductions.

Table 3.8

Changes in Farmer Net Cash Income and Payments Received from Federal Farm Programs Between 1982 and 1992

	1987 Average ($millions)		1987–1992 Average percentage change	
	Impact	Control	Impact	Control
Net farm cash returns	285	147	−20	−14
Government payments	27	8	−28	−64
Total net farm income	312	155	−21	−17

SOURCE: U.S. Department of Commerce, 1994.

There were several federal crop programs available to farmers during the drought that may have softened the impact of the drought on net farm income. For example, the "50/92" program for cotton allowed farmers to plant 50 percent of the acreage enrolled in the federal cotton program but receive payments as though he or she had planted 92 percent of the acres.[62] Table 3.8 suggests that farmers in the impact counties did take advantage of such programs. Farm program payments fell in both the impact and control counties between 1987 and 1992, but they fell much more slowly in the impact counties. Adding

---

[60]Net farm cash returns is the difference between the gross market value of agricultural products sold and total operating expenditures.

[61]Our analysis suggests that the drought had little effect on production and employment in the livestock and poultry sector. It could still be, however, that if livestock and poultry owners paid more for water that their net farm cash returns would fall.

[62]Farmers who could demonstrate that water cutbacks prohibited them from planting were also eligible for a special "0/92" program.

farm program payments to net farm cash returns somewhat reduces the difference between the impact and control counties in the percentage change in net farm income.

Cost and Availability of Credit

Water supply reductions or the threat of water supply reductions may affect the availability and cost of credit to the farm economy. Farmers depend on both short-term and long-term loans to finance their operations, and reductions in the availability of such loans may force them to scale back their operations. Lenders may be leery of making loans when water supplies are uncertain or may charge higher rates to compensate for increased risk of such loans. In this section we examine whether there were changes in the availability or cost of credit during the drought.

We were able to find no county-level data on the amount or cost of agricultural loans. This appears to be the case primarily because lenders regard lending information on such a localized level as highly proprietary. Nevertheless, we were able to assemble a qualitative picture of recent changes in the lending environment from review of the literature and interviews with a number of short- and long-term lenders in the San Joaquin Valley.

The four lenders we interviewed, by and large, said that decreased water availability in the San Joaquin Valley did not affect loan rates. Figure 3.8 provides some indirect evidence that water supply reductions did not affect rates much: the San Joaquin Valley accounts for a sizable proportion of California agriculture, but the spread between interest rates on short-term loans to farmers in the United States and California did not change much between 1988 and 1993.[63]

What apparently changed during the drought were the loan qualification requirements. Farmers were under increased pressure to demonstrate that they would be able to generate the income needed to pay back a loan. In the San Joaquin Valley, that meant demonstrating a reliable water supply.[64] The impetus for the more stringent lending requirements appears not to be the water supply reductions themselves, however, but the major losses among agricultural lenders nationwide during the mid-1980s (see Klonsky et al., 1994, for a good description of recent trends in the agricultural finance industry).

---

[63]What is more, several of the larger lenders we interviewed said they do not vary interest rates in California by region.

[64]Klonsky et al. (1994, p. 130) conclude that "what borrowers consider to be a 'credit crunch' in agriculture actually results from changes in the loan process and particularly credit analysis rather than changes in the availability or cost of funds."

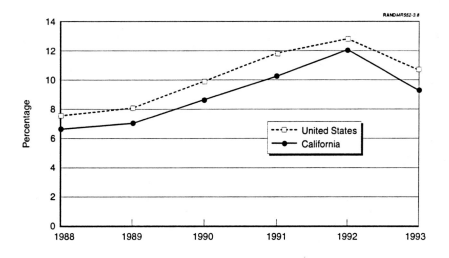

SOURCE: Personal communication with Guy Zimmerman, Federal Reserve Bank of San Francisco

Figure 3.8—Interest Rates on Short-Term Loans to Farmers

Reduced water availability thus means less credit, but the negative impact of this is not on farmers whose water supply reductions are fairly certain—they presumably need less because they are farming less. Rather, the negative effect is the inability to find credit of farmers who will likely receive water in the future but whose water supplies are uncertain.[65] Unfortunately, we have little empirical information on how often this is the case. In such cases, resolution of future water rights, even if it means lower deliveries, may mean increased credit availability because deliveries are more certain.

EVALUATION

Central to the debate on how water supply cutbacks in the San Joaquin Valley would affect agriculture is how flexible farmers can be in their response. If farmers are able to change their operations in numerous ways, the effect will be potentially less severe than if the only thing they can do is reduce the size of their operation. In Section 2, we suggested that a gradual shift from field crops to vegetables was a plausible response to permanent water cutbacks (see Table 2.6). The land value data presented here suggest that farmers indeed viewed at least part of the water supply cutbacks during the drought as permanent. The cropping pattern data are consistent with a steady shift to vegetables, induced by water supply cutbacks (see Table 3.9). Unfortunately, however, because the shift to vegetables was not well correlated with changes in water supply conditions, it may also be that the shift was due to factors other than water supply. For example, the higher water prices in the impact counties may make farmers there respond more quickly than farmers in the

---

[65]Such uncertainty clearly affects long-term loans, but it may even affect very short-term loans. Short-term lenders may not be willing to make the investment in building a relationship with a farmer if that farmer is not going to be in business over the long run.

control counties to market conditions such as crop prices. Similar conclusions apply to shifts from low- to high-value field crops.

Crop fallowing is the widely expected response to temporary water cutbacks, and without a doubt, we saw significant fallowing of field crops. We also saw that farmers in the impact counties fallowed both low- and high-value field crops. Farmers can also modify their irrigation practices in response to water supply reductions. Changes in irrigation technology are likely in the response to permanent water cutbacks, and changes in irrigation management and deficit irrigation are likely in response to temporary cutbacks. As summarized in Table 3.9, we found no evidence that farmers stressed their crops enough to reduce yields during the drought. Surprisingly, we also found no evidence in lasting improvement in irrigation efficiency, but our data are weak.

We argued in Section 2 that water supply reductions unambiguously make farmers worse off. This was indeed the case during the drought. Farmers saw lower profits, and land values declined. The effect on agricultural employment, both on and off the farm, is less clear. Shifts to vegetables and more advanced irrigation practices may increase demand for certain inputs and types of labor over the long run. Farmers may also hold on to their labor supply during temporary cutbacks. We found no clear evidence that the water supply reductions during the drought caused a fall in agricultural employment, although under some assumptions it is possible to conclude that on-farm crop production did fall. Our analysis, therefore, does suggest that farmers may respond in such a way that the employment effects of water supply cutbacks in the range observed during the drought may not be great.

Table 3.9

Summary of Effects of Water Supply Reductions During the 1986–1992 Drought in the Impact Counties

Measure of Activity	Summary
Water use	
Surface water	Fell dramatically in impact counties, much less in controls
Total water	Fell 15% 1987–1989 to '91 and 7% 1987–1989 to '92 in impact counties; unchanged in controls.
Crop pattern	
Acreage harvested	Fallowed both low- and high-value field crops
Crop mix	Weak evidence of shift from field crops to vegetables and from low- to high-value field crops
Irrigation practices	
Crop yield	No evidence of reduced yield
Irrigation management	Water use data consistent with improvement in irrigation management in 1991, but deficit irrigation also a possible explanation
Irrigation technology	No indirect evidence of changes in irrigation technology
Value of production	
Crop production	Reduced field crop production reduces overall crop production value 6 to 7% between 1987–1989 and '91, or '92; half of this decline offset by increased vegetable production
Livestock and poultry production	Somewhat faster growth in the impact counties relative to controls
Employment	
Crop production	No clear evidence of any effect, but 5% reduction in on-farm crop production possible
Livestock and poultry production	No negative effect
Countywide	No negative effect detectable
Farmer profit	Approximately a 4% decline
Land values	Fell approximately $125 per acre in impact counties relative to controls
Access to credit	No change in loan rates, reduced access to credit for those with uncertain water supplies

4. MODEL PREDICTIONS OF THE IMPACT OF REDUCED WATER SUPPLIES

This section examines two economic models that are commonly used to predict the impact of water supply cutbacks. Even though these models are widely used by regulatory agencies, surprisingly little is known about the realism of their assumptions and the accuracy of their predictions (U.S. EPA, 1994; California State Water Resources Control Board, 1994). We will describe the models and compare their predictions of the effects of water supply cutbacks with the empirical findings in Section 3. This will allow us to better understand how the assumptions behind the models correspond to behaviors we might expect from theory and how well their predictions correspond to reality.

In the remainder of this section, we first describe the two models. We then report the model estimates of the impact of water cutbacks during the drought in Fresno and Kern Counties and compare them to findings from Section 3. We finally discuss the implications of the findings.

MODELS USED TO PREDICT IMPACTS OF THE DROUGHT

The models evaluated in this section are the rationing model and the Central Valley Production Model (CVPM). Each of the models is described in turn.

The Rationing Model

The rationing model was initially proposed by researchers at the University of California at Berkeley (Zilberman, Howitt, and Sunding, 1993) and then refined during EPAs analysis of the impact of its proposed Bay/Delta water quality standards. It predicts changes in crop acreage and crop revenue (gross and net) caused by changes in water supply. The rationing model divides the Central Valley into 21 regions. It assumes that farmers respond to water cutbacks by fallowing the "lowest value" crops in each region until the amount of water saved equals the cutback. Crops are ranked according to net revenue per acre-foot of water applied to the crop. Regional estimates of the net revenue (crop revenue less crop production costs) per acre-foot of applied water are made using Bureau of Reclamation crop budgets.[66] The model fallows the lowest-value crops across the entire region—in effect assuming that farmers can and do transfer enough water within the region to allow this to happen.[67]

Table 4.1 lists the average net and gross crop revenues per acre-foot that are used in the rationing model run below. The rationing model will first fallow field crops, and only

---

[66]Groundwater costs are not included in the model. This affects net revenue (not gross revenue) and needs to be estimated separately.

[67]For example, if Farmer A grows only fruits and nuts and Farmer B grows only low-value field crops but they both face the same proportionate cutbacks in water supply, then Farmer B will sell water to Farmer A. Farmer B will fallow his or her entire acreage before Farmer A fallows anything.

when all field crops in a particular region are exhausted will it move to fruits, nuts, and vegetables. The last field crops to be fallowed will be alfalfa, dry beans, sugar beets, and cotton, which are the *high-value* field crops defined in Section 3. Note that the ordering of the crops does not change much if gross revenue is used as the criterion for fallowing crops instead of net revenue. Thus, a rationing model that fallows crops based on their gross revenue per acre-foot will yield very similar results to one that uses net revenue. Some researchers have used gross revenue (e.g. Zilberman, Howitt, and Sunding, 1993) because gross revenue numbers are easy to obtain and usually perceived to be more accurate than the net revenue numbers.

Table 4.1

Average Net and Gross Revenue Per Unit of Applied Water for Crops Commonly Grown in the San Joaquin Valley
($/acre-foot)

Crop	Net Revenue per acre-foot	Gross Revenue per acre-foot
Field crops		
Rice	4	76
Irrigated pasture	14	40
Corn	81	166
Wheat	98	210
Alfalfa	100	167
Dry beans	121	247
Sugar beets	142	269
Cotton	168	338
Fruit and nuts		
Raisin grapes	297	567
Walnuts	338	575
Almonds	348	560
Citrus	943	1,296
Vegetables		
Processing tomatoes	298	530
Melon	311	1,279
Onions	719	2,613
Potatoes	809	1,762

SOURCE: U.S. Department of the Interior, 1994.

The rationing model places a number of restrictive assumptions on farmer responses to water supply reductions and limitations on the factors that farmers consider in making decisions. The model makes the following assumptions:

- Farmers do not shift into low-water-using crops or to more labor- and capital-intensive crops as the shadow price of water rises in response to water cutbacks.
- Farmers do not consider crop rotation requirements, crop commodity programs, or variation in growing and market risk across crops when deciding which crops to fallow.

- Farmers do not increase farm irrigation efficiency during a water shortage either by adopting new irrigation technologies or improving irrigation management practices.
- Crop prices remain fixed during water shortages, and crop acreage cannot be changed in response to crop price changes.

The rationing model also does not predict how groundwater use will change in response to change in surface water use. Rather, the change in total water use must be input into the model.[68]

The rationing model is not based on standard economic theory of farmer decisionmaking. Economic theory suggests that farmers will equalize the marginal net revenue per acre-foot across crops. Therefore, assuming that farmers fallow acreage based on the average net revenue per acre-foot may not be realistic. Also, the discussion in Section 2 suggests that many of the responses and considerations listed above may be relevant even in the short run, and there is no obvious theoretical grounds to exclude them.

A key advantage of the rationing model is its simplicity. How the model works is transparent, and once data on cropping patterns and estimates of net revenue per acre-foot are in hand, it can be quickly run. Some of the assumptions of the rationing model may be realistic for responses to temporary water supply cutbacks and short-run responses to permanent cutbacks, but there is little theoretical reason to expect it to produce even first-order estimates of the actual effects of water supply cutbacks.

The Central Valley Production Model

The CVPM was developed by researchers U.C. Davis and the California Department of Water Resources (U.S. Department of the Interior, 1994). The CVPM assumes that farmers act to maximize farm profits subject to water (and other resource) constraints and market conditions.

The CVPM places far fewer restrictions on farmer response to water cutbacks than the rationing model does. The CVPM allows farmers to shift between crops depending on how the water cutback affects the shadow price of water. Farmers may shift to low-water-using crops and increase irrigation efficiency (by investments in irrigation technology). Profit maximization is what drives all these decisions. The CVPM assumes that farmers can distinguish between low- and high-yielding acreage and allows the marginal cost of production to vary within each of its 21 regions. It also includes estimates of crop demand elasticities so that farmers can take advantage of increased prices for those crops that are taken out of production but risk lower prices when there are increases in production.

In contrast to the rationing model, the CVPM assumes that high- and low-value crops are equally profitable per acre-foot of water applied at the margin. This means that the amount of each crop fallowed depends on the relative slopes of the crop's marginal value of

---

[68]Extensions of the rationing model might compare net revenue per acre-foot with groundwater cost to determine changes in groundwater use, but such extensions are not examined here.

water functions. The CVPM may thus predict changes in both crops with high average net revenue per acre-foot and low average net revenue per acre-foot. The rationing model, in contrast would restrict changes to crops with low net revenue per acre-foot.

The CVPM may predict increases in the acreage of some crops in response to water cutbacks. There are at least two circumstances where acreage of some crops may increase. The first is when a water shortage in one area causes a decrease in the production of a crop, an increase in its price, and farmers in other areas respond to the price increase by increasing acreage of the crop. The second is when the water shortage causes an increase in the shadow cost of water and farmers respond by increasing the acreage of low-water-using crops or crops for which irrigation efficiency can be easily improved.

Like the rationing model, the amount of surface water available is an input into the CVPM, but unlike the rationing model, CVPM determines internally the amount of groundwater pumped by farmers. The amount pumped is determined by the cost of groundwater, which is a function of the specified depth-to-groundwater in each region. In contrast, the amount of groundwater used in the rationing model must be specified externally—which is difficult to determine either in advance of water cutbacks, because it is up to farmers, or *ex post* because the data on groundwater pumping are often poor.

The advantage of CVPM over the rationing model is that it allows farmers a rich set of responses to water cutbacks. Like the rationing model, it does not incorporate variation in crop or market risk across crops, rotation requirements, or the requirements of crop commodity programs, but it nevertheless permits much more flexible response to water cutbacks than the rationing model does.

The disadvantage of CVPM is that it is a complex non-linear programming model that is hard to run, and its workings are difficult to understand. It uses a cleaver method to equate the marginal net revenue per acre-foot for all crops, but this may severely limit its predictive power because too many parameters are estimated using limited data (the model is overfit). Also, the model contains constraints that limit the changes in acreage on a crop- and region-specific basis that can result from a change in water supply. It is unclear what these constraints are based on, and they may have an important influence on a particular simulation.

MODEL PREDICTIONS OF THE 1991 DROUGHT IMPACT

In this section, we use the rationing model and the CVPM to predict the impact of water supply cutbacks in Fresno and Kern Counties between 1987–1989 and 1991. This was the period with the most dramatic decline in surface water supplies during the drought. We first describe the assumptions used to run the models and then discuss the results.

Assumptions Used in Simulations

For the rationing model, we compare cropping pattern and crop revenue in 1988 in Fresno and Kern Counties with the acreage and revenue predicted when total water use is reduced 15 percent. This 15 percent reduction matches the average reduction reported in

Section 3 for Fresno and Kern Counties between 1987–1989 and 1991. The reduction is made from the CVPET estimate of total water use in 1988.[69]

For the CVPM, we compare cropping pattern and crop revenue predicted by the model using 1988 surface water supplies in Fresno and Kern with cropping pattern and crop revenue predicted when surface water supply is decreased 62 percent in Fresno and 78 percent in Kern. These drops match the declines reported in Section 3 between 1987–1989 and 1991. Groundwater use is determined by the model itself, thus total water use in the base year, or 1991, will not necessarily equal our estimate of total water use in 1988, or 1991, nor will the change in total water use necessarily equal that observed.

Results

Table 4.2 presents the average percentage changes in acreage, gross crop revenue, and water use predicted by the rationing model and the CVPM in response to the declines in total water and surface water, respectively, observed between 1987–1989 and 1991.[70]

Water Use. By construction, total water use falls the same in the rationing model as observed in Fresno and Kern Counties between 1987–1989 and 1991.[71] CVPM predicts that water use falls an average of 14 percent, quite close to the observed change in water use. This means that the CVPM predicted the increase in groundwater pumping quite accurately.

Acreage Harvested. Neither model predicts a shift to vegetables. The data in Section 3 suggests that there may have been a shift to vegetables (vegetable acreage grew 9 percent faster in the impact counties than in the control counties between 1987–1989 and 1991), but the evidence was weak and the shift could have been due to factors other than the drought. Thus, the model predictions may be accurately predicting that water supply cutbacks produced no shift to vegetables during the drought. Both models also predict no change in fruits and nuts, which is consistent with our interpretation of the empirical data.

---

[69]The CVPET estimates of total water use for Fresno and Kern Counties are 6 to 7 percent greater than the derived water use estimate for 1988. Thus, from the perspective of the rationing model, which, like the derived water use numbers, determines total water use by crop-specific water-use coefficients, there is "excess" water in the system in 1988. The 15 percent water reduction used in the simulations must then go through this excess water before there is any effect on cropping pattern. We think it makes most sense to do the simulations this way because the water data suggest that there really was excess water from the perspective of the rationing model in 1988. There may be excess water because the implied water use numbers assume unrealistically high irrigation efficiencies or that farmers are intentionally putting on more water than necessary to recharge groundwater.

[70]Employment changes are not directly predicted by the models nor are effects on input suppliers and output processors.

[71]Because surface water is the input to CVPM, the entire discrepancy in total water use is due to difference in groundwater use.

Table 4.2

Model Predictions of Impacts of Water Supply Cutbacks in Fresno and Kern Counties

	1987–1989 Baseline	Average Percentage Change From 1987–1989 Baseline to 1991		
		Observed[a]	Rationing Model	CVPM
Total water use (1000s of acre-feet)	3,318	−15	−15	−14
Acreage (1000s)	1,028	−14	−9	−12
Field crops	623	−22	−12	−20
Fruits and nuts	265	0	0	0
Vegetables	138	9	0	−2
Type of field crop				
Low value	161	−35	−38	−18
High value	465	−22	−3	−21
Gross crop revenue ($millions)	1,467	−3 to −7	−2	−11

[a]Our interpretation of the observed effect of water supply cutbacks is based on the information presented in Tables 3.4 and 3.6. Estimates are usually based on differences in changes between the impact and control counties.

The CVPM more accurately predicts the reduction of field crops and total acreage than the rationing model. Likewise, the CVPM more accurately captures the reductions in low- and high-value field crops. The rationing model predicts that fallowing will be almost exclusively restricted to low-value field crops. The CVPM, in contrast, predicts substantial percentage reductions for both low- and high-value field crops, and actually predicts a greater reduction in the acreage of high-value than low-value field crops.[72] Indeed, the empirical data suggest that more acres of high-value field crops were fallowed than low-value field crops between 1987–1989 and 1991.

Farm Revenue. As expected, because it concentrates fallowing on the low-value field crops, the rationing model predicts a lower reduction in gross crop revenue than the CVPM does. The observed decline in crop revenue lies between the rationing model's predicted 2 percent decline and the CVPM's predicted 11 percent decline. Recall from Section 3 that if, as the two models suggest, the shift to vegetables in the impact counties was due to factors other than water supply cutbacks, the actual fall in crop revenue due to water supply cutbacks was 6 or 7 percent.

EVALUATION

Once groundwater use is specified, the rationing model provides a first-order approximation of some of the responses to the combination of temporary and permanent water supply cutbacks during the 1986–1992 drought. The predicted change in total

---

[72]This is because the acreage in high-value field crops at baseline is much greater than that in low-value field crops.

acreage harvested is not too far off, and the model accurately predicts that fallowing was limited to field crops. The model does not do well at a more detailed level, however. The assumption that farmers will fallow crops according to their net revenue per acre-foot appears incorrect. And, the overconcentration on low-value field crops means that the change in gross crop revenue predicted by the rationing model is too low. The rationing model could conceivably do better if Kern and Fresno were broken down into smaller regions.[73] But, this would require more-detailed water use and cropping data and more assumptions on water trading among regions.[74]

An important factor limiting the usefulness of the rationing model is the requirement that the amount of groundwater pumping be specified. Most agricultural areas in California have access to groundwater and predicting how a farmer's groundwater use will respond to surface water cutbacks is a difficult job.

The CVPM does a reasonably good job of predicting the impact of water supply cutbacks on cropping pattern. An ongoing concern about the CVPM, however, is the role that constraints play in the model. These constraints may prohibit the CVPM from showing a shift to vegetables, and may otherwise inappropriately limit the responses farmers can make to water supply reductions.

---

[73]If regions were smaller and water cutbacks were concentrated in relatively few regions, the rationing model would likely predict more fallowing of high-value field crops than is currently the case.

[74]Extensive trading could mitigate the effect of running the model with smaller regions.

5. EVALUATION AND POLICY IMPLICATIONS

New regulations on water quality in the San Francisco Bay/Delta will likely mean permanent reductions in surface water deliveries.[75] In the short run, farmers may offset part or all of these declines with increases in groundwater pumping. Over time, however, groundwater levels will decline, increasing the cost of groundwater and reducing the amount of groundwater pumped. It thus seems likely that a large fraction, if not all, of the reduction in surface water supplies will ultimately be reflected in a reduction in total water use.

Our review of theory, past empirical work, the effect of the 1986–1992 drought, and models used to predict the effect of water supply reductions offers the following lessons on the impact of permanent water supply reductions on the San Joaquin Valley.

POOR DATA ON WATER USE MAKES ANALYSIS DIFFICULT

The lack of reliable data on groundwater pumping in the San Joaquin Valley makes it very difficult to determine how farmers respond to water supply cutbacks. The lack of good data makes it difficult to use past experience to predict how future regulatory water reductions will affect agriculture. Better measurement of actual water use is needed.

IMPROVEMENTS IN IRRIGATION EFFICIENCY IN RESPONSE TO PERMANENT REDUCTIONS LIKELY OVER THE LONG RUN

Strong theoretical arguments and strong empirical evidence suggest that farmers will shift to sprinkler and drip irrigation in response to permanent water supply reductions. We did not have good data with which to examine the change in irrigation technology and management during the drought, but others have concluded that there were widespread improvements in irrigation management, although whether there was much shift to sprinkler and drip is under dispute. One expects to see such changes occur only gradually and in response to permanent, not temporary, cutbacks. Improvements in irrigation efficiency should allow farmers to stretch water supplies further.

THE EFFECT OF WATER SUPPLY REDUCTIONS ON CROP MIX REMAINS UNCERTAIN

Theory suggests that an increase in the shadow price of water may induce farmers to shift to crops that need less water or that have higher labor or capital intensities. There has been some empirical support for such changes in the past, but the evidence is limited. There was an increase in vegetable acreage during the drought in the impact counties, but uncertainty remains about whether this was due to the water supply cutbacks or to other factors unrelated to water supply. Similar uncertainties remain on the shift from low- to

---

[75]There will still be variation across years, but the mean around which annual deliveries fluctuate will be lower.

high-value field crops. Any such shifts would be limited by the effect of increased production of, say, vegetables, on vegetable prices.

EMPLOYMENT EFFECTS OF LONG-TERM SUPPLY REDUCTIONS AMBIGUOUS

The effects on long-term water supply reductions on agricultural employment are ambiguous in theory. Shifts to more labor-intensive crops or irrigation systems may offset any change, say, in the amount of acreage farmed.

The lack of any clear reduction in agricultural employment during the drought is surprising. Not only was there no discernible effect on overall employment in Fresno and Kern Counties, but there was not even strong evidence that on-farm crop production employment fell. It may be that our employment data did not capture significant effects among seasonal or undocumented workers, that farmers held on to employees in order to maintain their labor force for the future during what they perceived to be, in part, temporary water cutbacks, or that shifts to vegetables unrelated to the water supply cutbacks masked employment reductions during the drought. More work and better data are needed to sort out these possibilities, but for the time being, the impact of even substantial water cutbacks on agricultural employment remains uncertain.

FARMERS AND LANDOWNERS ARE ADVERSELY AFFECTED BY WATER REDUCTIONS

Water supply reductions, whether they are temporary or permanent, will negatively affect farmers. Reductions represent constraints on farm operations, and farmers can be no better off with them than without them. Agricultural land values represent the expected long-run profitability of farming and thus will also likely fall in response to water supply cutbacks. Data suggest that San Joaquin farmers and landlords were adversely affected during the drought: Farmer profit fell as did relative land prices in the counties that faced the largest water supply cutbacks.

RATIONING MODEL IS A LIMITED TOOL FOR POLICY ANALYSIS, CENTRAL VALLEY PRODUCTION MODEL NEEDS TO BE BETTER UNDERSTOOD

The rationing model provides first-order approximations for at least some farmer responses to the type of water supply reductions that occurred during the drought. However, it does not do well at a more detailed level, and its central underlying assumption—that farmers fallow crops with the lowest net revenue per acre-foot of water applied—appears incorrect. Its inability to predict changes in groundwater pumping is also a major drawback. The rationing model's simplicity and ease of use is tempered by this limitation and the inaccuracy of many of its predictions.

The CVPM appears to be a better choice for policy analysis, but the appropriateness of its many constraints needs to be better understood. Its complexity and intense data requirements temper its apparent accuracy in predicting responses to the 1986–1992 drought.

ONGOING EVALUATION OF FARMER RESPONSE TO WATER SUPPLY REDUCTIONS NEEDED

Many uncertainties remain about how farmers will respond to permanent water supply cutbacks, and ongoing study of farmer response to the water supply reductions is needed. The empirical analysis in this report suffered from a limited number of counties and a limited number of years of data. More counties and data farther back in time would certainly help to resolve some of the uncertainties we faced in interpreting the numbers, and further research should be done. A more promising approach may be to analyze response to water supply cutbacks at the farm level, and it would also seem productive to use cross sectional data to further examine the effect on farm prices of water prices and availability. It is difficult and expensive to collect this micro-level data, but the resulting findings could be much more definitive.

Better information on the effects of water supply reductions on agriculture will allow policymakers to revisit decisions to reallocate water from agriculture to the environment with more accurate information on the costs and benefits.

Appendix

A. SUPPLEMENTAL DATA ON IMPACT AND CONTROL COUNTIES

This appendix presents additional information on the impact and control counties. Table A.1 contains data on 1985 county population, total personal income, and employment. Table A.2 presents both the Central Valley Project Environmental Team (CVPET) and derived estimates of the change in surface water, groundwater, and total water use in the individual counties between 1985 and 1992.

As can be seen in Table A.2, average derived total water use in 1985 and 1986 is 5 to 8 percent below the CVPET estimates for each of the counties. One factor causing this difference is that the derived water estimates are based on acreage harvested rather than on acreage planted. Acreage harvested in the San Joaquin Valley is roughly 5 percent less than acreage harvested (U.S. Department of Commerce, 1994) biasing the derived water use estimates down.

Table A.1

Population, Personal Income, and Employment in the Impact and Control Counties in 1985

	Impact Counties		Control Counties		
	Fresno	Kern	Merced	Stanislaus	San Joaquin
Population (millions)[a]	0.572	0.473	0.158	0.301	0.408
Total personal income ($billions)[b]	7.391	6.180	1.783	3.913	5.393
Percentage from agriculture	11	9	12	6	5
Total wage and salary employment[c]	254,700	206,100	62,775	123,800	162,600
Percentage in agriculture[d]	21	13	16	11	10
Total farm wage & salary employment (1000s)[e]	54,210	27,535	10,147	13,626	15,670
Crop production (%)	94	96	82	74	88
Livestock and poultry (%)	6	4	18	26	12
Seasonal wage and salary workers[f] (percentage of agricultural workers)	66	53	73	64	73

[a]California Department of Finance, 1993, p. 3.

[b]U.S. Department of Commerce, 1994. Agriculture income includes farm income and income from agricultural services.

[c]California Employment Development Department, 1993.

[d]Includes employment on farming operations and in firms that provide agricultural services. Does not include employment in all firms that supply farm inputs, such as seed and pesticide suppliers, or firms that process farm products such as canners.

[e]Special reports generated by California Employment Development Department, 1994; includes agricultural services.

[f]California Employment Development Department, 1987.

Table A.2

Change in Water Use During the Drought by County

	Impact Counties		Control Counties		
	Fresno	Kern	Merced	Stanislaus	San Joaquin
CVPET total water use					
1985–1986 (1000s of acre-feet)	3,894	2,743	1,698	1,554	1,128
1985–1986 to 1987–1989 (% change)	1	–1	–3	0	1
1987–1989 to 1991 (% change)	–14	–16	2	–1	4
1987–1989 to 1992 (% change)	–6	–8	5	4	5
CVPET surface water diversions					
1985–1986 (1000s of acre-feet)	2,578	1,722	1,098	1,047	651
1985–1986 to 1987–1989 (% change)	–15	–21	–13	–6	5
1987–1989 to 1991 (% change)	–62	–78	–21	–12	3
1987–1989 to 1992 (% change)	–55	–50	–27	–14	3
CVPET groundwater water use					
1985–1986 (1000s of acre-feet)	1,315	1,020	600	507	477
1985–1986 to 1987–1989 (% change)	32	32	17	11	–5
1987–1989 to 1991 (% change)	48	46	34	19	7
1987–1989 to 1992 (% change)	57	34	49	35	7
Derived total water use					
1985–1986 (1000s of acre-feet)	3,649	2,594	1,588	1,444	1,041
1985–1986 to 1987–1989 (% change)	3	–2	8	4	–1
1987–1989 to 1991 (% change)	–6	–8	4	6	–3
1987–1989 to 1992 (% change)	–6	–6	1	4	–4
Derived groundwater use					
1985–1986 (1000s of acre-feet)	1,071	871	490	398	391
1985–1986 to 1987–1989 (% change)	46	38	56	29	–14
1987–1989 to 1991 (% change)	73	70	35	40	–16
1987–1989 to 1992 (% change)	62	43	36	39	–20

B. THE DETERMINANTS OF AGRICULTURAL LAND VALUES

This appendix provides some background on the effect of water availability and other factors on agricultural land values in the San Joaquin Valley.

There is little doubt that the availability of water is a key determinant of land value in the San Joaquin Valley. Water increases crop productivity and thus the value of land for crop production. Table B.1 illustrates the link between land value and water availability by comparing the value of irrigated pasture in the San Joaquin Valley with the value of non-irrigated pasture. Between 1982 and 1991, the value of irrigated pasture was $1,227 per acre greater than the value of non-irrigated pasture, and there was not much deviation of the annual differences from this average. The consistent difference between irrigated and non-irrigated pasture suggests that water supply increases the value of San Joaquin Valley pasture approximately $1,200 per acre (in 1985 dollars).[76]

Table B.1

Comparison Between the Value of Irrigated and Non-Irrigated Pasture Land in the San Joaquin Valley
(1985 dollars)

Year	Irrigated Pasture	Non-Irrigated Pasture	Difference in Value
1982	$2,725	$1,475	$1,250
1983	2,738	1,310	1,429
1984	2,490	1,251	1,239
1985	2,300	1,050	1,250
1986	1,949	877	1,072
1987	1,936	803	1,133
1988	2,089	817	1,272
1989	2,045	914	1,131
1990	2,083	875	1,208
1991	2,167	883	1,284
Period Average	2,252	1,025	1,227

SOURCE: California Agricultural Statistics Service (1994). Data are based on a survey of farmers that asked the value of farmland excluding the value of buildings.

Of course it is not just the availability and price of water that determines land values. The amount of investment in land improvements is also important. Land devoted to pasture is usually the least expensive farmland because few improvements are needed to grow pasture. Land devoted to vegetables and orchard crops is usually more expensive, in part because investment is required to improve the land before orchard and vegetable production is possible. (This investment may include the cost of leveling, furrowing, and other land

[76]It should be noted that these data are only suggestive of the value of water supply since the quality of pasture land with water supply may differ from the quality of pasture land without water supply.

improvements as well as the cost of tree plantings and other land investments). As an illustration, the price of vegetable crop land in the San Joaquin Valley was approximately 55 percent greater than irrigated pasture land between 1982 and 1991, and the average price of almond orchard land was over 150 percent greater than the price of irrigated pasture land (California Agricultural Statistics Service, 1994).

Economywide economic cycles also affect land values. Between 1982 and 1991, the price of farmland across the United States, in California, and in the San Joaquin Valley first declined significantly and then increased (see Figure B.1). The decline and subsequent rise was larger in the San Joaquin Valley than in California or in the United States as a whole. The price cycle may have been caused by changing expectations about crop prices (or other variables that affect the expected crop land returns).

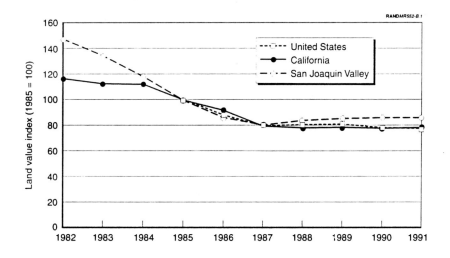

SOURCE: California Agricultural Statistics, 1994. U.S. figures not available for 1982–1984.

Figure B.1—Index of Agricultural Land Values in the United States, California, and the San Joaquin Valley

REFERENCES

Archibald, Sandra, *Economic Profile of Agriculture in the West Side of the San Joaquin Valley*, Food Research Institute, Stanford University, June 1990.

Archibald, Sandra O., and Thomas E. Kuhnie, *An Economic Analysis of Water Availability in California Central Valley Agriculture: Phase III Draft Report*, Center for Economic Policy Research, Stanford University, 1994.

Archibald, Sandra O., Thomas E. Kuhnie, Robin Marsh, Mary Renwick, and Barton Thompson, Jr., *An Economic Analysis of Water Availability in California Central Valley Agriculture: Phase II Draft Report*, Center for Economic Policy Research, Stanford University, February 14, 1992.

California Agricultural Statistics Service, 1994.

California Department of Finance, *California Statistical Abstract, 1993*, Sacramento, California, 1993.

California Department of Water Resources (CDWR), *Crop Water Use in California*, Bulletin 113-4, Sacramento, California, April 1986.

California Department of Water Resources (CDWR), *California's Continuing Drought, 1987–1991: A Summary of Impacts and Conditions As of December 1, 1991,* Sacramento, California, December 1991.

California Department of Water Resources (CDWR), Division of Planning, Statewide Planning Branch, *Agricultural Water Use Biennial Report*, May 1993.

California Department of Water Resources (CDWR), *California Water Plan Update*, Bulletin 160-93, Sacramento, California, 1994.

California Employment Development Department, Job Service Division, *Report 881-M: Agricultural Employment Estimates, 1985 and 1986*, February 1987.

California Employment Development Department, Special Projects Unit, *California Drought Bulletin,* monthly issues, April through September 1991.

California Employment Development Department, Labor Market Information Division, *Annual Planning Information*, for various counties, 1993.

California State Water Resources Control Board, *Draft Environmental Report for Implementation of the 1995 Bay-Delta Water Quality Control Plan,* November 1997.

Caswell, Margaret, and David Zilberman, "The Choices of Irrigation Technologies in California," *American Journal of Agricultural Economics*, 1985, pp. 224-234.

Dixon, Lloyd S., Nancy Y. Moore, and Susan W. Schechter, *California's 1991 Drought Water Bank: Economic Impacts in the Selling Regions*, Santa Monica, Calif.: RAND, MR-301-CDWR/RC, 1993.

Dixon, Lloyd S., *Models of Groundwater Extraction with an Examination of Agricultural Water Use in Kern County, California*, Ph.D. Dissertation, U.C. Berkeley, 1988.

Klonsky, Karen, Steven C. Blank, Robert C. Thompson, Jr., Thomas W. Hazlett, and Lawrence Shepard, "The Supply of Credit to Agriculture," in *Financing Agriculture in California's New Risk Environment*, Steven C. Blank, ed., Agricultural Issues Center, University of California, Davis, California, March 1994.

Mamer, John, and Alexa Wilkie, *Seasonal Labor in California Agriculture: Labor Inputs for California Crops*, California Employment Development Department, California Agricultural Studies, 90-6, December 1990.

Moore, Michael R., Noel R. Gollehon, and Marc B. Carey, "Multicrop Production Decisions in Western Irrigated Agriculture: The Role of Water Price," *American Journal of Agricultural Economics,* Vol. 75, pp. 859–874, November 1994.

Negri, Donald H., and Douglas H. Brooks, "Determinants of Irrigation Technology Choice," *Western Journal of Agricultural Economics,* 15(2), pp. 213–223, 1990.

Northwest Economic Associates, *Economic Impacts of the 1991 California Drought on San Joaquin Valley Agriculture and Related Industries*, Vancouver, Washington, March 16, 1992.

U.S. Department of Agriculture, *National Agricultural Statistics Service, Agricultural Prices: 1993 Summary,* Pr 1-3(94), Washington, D.C., July 1994.

U.S. Department of Commerce, Bureau of the Census, *United States Statistical Abstract: 1986,* Washington, D.C., 1986.

U.S. Department of Commerce, Bureau of the Census, *1992 Census of Agriculture*, Volume 1, Part 5, California State and County Data, Washington, D.C., September 1994.

U.S. Department of Commerce, Bureau of the Census, *1987 Census of Agriculture*, Volume 1, Part 5, California State and County Data, Washington, D.C., May, 1989.

U.S. Department of Commerce, Bureau of Economic Analysis, Regional Economic Information System, May 1994.

U.S. Department of the Interior, Bureau of Reclamation, *Central Valley Production Model and Supporting Data*, Sacramento, Calif., 1994.

U.S. Environmental Protection Agency, Region 9, *Regulatory Impact Assessment of the Final Water Quality Standards for the San Francisco Bay/Delta and Critical Habitat Requirements for the Delta Smelt,* Internal Review Draft, San Francisco, Calif., October 5, 1994 (with technical assistance from Jones & Stokes Associates, Inc., Sacramento, Calif.).

U.S. House of Representatives, Committee on Natural Resources, *Taking From the Taxpayer: Public Subsidies for Natural Resource Development,* U.S. Government Printing Office, Washington D.C., August 1994.

Zilberman, David, Ariel Dinar, Neal MacDougall, Madhu Khanna, Cheryl Brown, and Frederico Castillo, *Individual, Organizational, and Institutional Responses to the Drought: The Case of California Agriculture,* Department of Agricultural and Resource Economics, U.C. Berkeley, 1994.

Zilberman, David, Richard Howitt, and David Sunding, *Economic Impacts of Water Quality Regulations in the San Francisco Bay and Delta*, Berkeley, Calif.: Western Consortium for Public Health, May 1993.